LIVESTOCK GUARDIANS

STOREY'S WORKING ANIMALS

LIVESTOCK GUARDIANS

USING DOGS, DONKEYS and LLAMAS TO PROTECT YOUR HERD

Janet Vorwald Dohner

Storey Publishing

The mission of Storey Publishing is to serve our customers by
publishing practical information that encourages
personal independence in harmony with the environment.

Edited by Lisa Hiley and Deborah Burns

Art direction by Alethea Morrison and Cynthia McFarland
Text design and production by Vicky Vaughn Design
Cover design by Alethea Morrison

Cover and chapter opener illustrations by John MacDonald
Letterpress typography and borders by Yee Haw Industries
Illustrations by © Elayne Sears, except for pages 23, 30 bottom, 31, 32 top two, and 33 by
  Brigita Fuhrmann
Maps by Ilona Sherratt
Back cover photographs by © Bruce N. Gilbert/Painet, Inc.: middle; © Cat Urbigkit: top and
  bottom
Interior photography credits appear on page 219

Indexed by Susan Olason, Indexes and Knowledge Maps

© 2007 by Janet Vorwald Dohner

All rights reserved. No part of this book may be reproduced without written permission from the publisher, except by a reviewer who may quote brief passages or reproduce illustrations in a review with appropriate credits; nor may any part of this book be reproduced, stored in a retrieval system, or transmitted in any form or by any means — electronic, mechanical, photocopying, recording, or other — without written permission from the publisher.

The information in this book is true and complete to the best of our knowledge. All recommendations are made without guarantee on the part of the author or Storey Publishing. The author and publisher disclaim any liability in connection with the use of this information. For additional information, please contact Storey Publishing, 210 MASS MoCA Way, North Adams, MA 01247.

Storey books are available for special premium and promotional uses and for customized editions. For further information, please call 1-800-793-9396.

Printed in the United States by Versa Press
10 9 8 7 6 5 4 3 2 1

LIBRARY OF CONGRESS CATALOGING-IN-PUBLICATION DATA

Dohner, Janet Vorwald, 1951–
  Livestock guardians / Janet Vorwald Dohner.
    p. cm. — (Storey's working animal series)
  Includes index.
  ISBN 978-1-58017-695-8 (pbk. : alk. paper); ISBN 978-1-58017-696-5 (hardcover : alk. paper)
  1. Livestock—Predators of—Control. 2. Working animals. 3. Livestock protection dogs.
  I. Title. II. Series.
SF810.5.D64 2008
636.08'39—dc22

2007024811

ACKNOWLEDGMENTS

I MUST THANK the many individuals and groups who study, promote, and work with these animals. The people who breed these animals are incredibly dedicated. Many of you have shared your knowledge and insights with me, and I am most grateful. We must continue to communicate and share our information and observations. Since working with guardian animals is more art than science, sharing our personal experiences is invaluable. We all learn so much from each other, and we still have much to learn. Please continue to send information or suggestions to make future editions of this book even more helpful to the person who wishes to use a guardian animal.

Thank you to my family, friends, and professional colleagues who support this curious passion. The members of my particular dog breed community are incredibly selfless in sharing everything. In addition, grateful thanks to the folks at Storey Publishing, who remain committed to environmentally healthy methods of farming.

Finally, I have learned the most from the dogs themselves. I continue to marvel at their intelligence, their ability to reason, their essential good natures, their powerful instincts, their loyalty and devotion, their gentleness and ferocity, and especially their individual differences. Through the years, these special dogs live in my heart: Bear, Kizzie, Kelebek, Peshi, and Fistik. Good dogs.

CONTENTS

Foreword by D. Phillip Sponenberg, DVM, PhD	viii
Preface	x
1 THE PROBLEM OF PREDATION	2
2 METHODS OF PREDATOR CONTROL	23
3 LIVESTOCK GUARD DOGS	39
4 SELECTING AND TRAINING YOUR DOG	50
5 TAKING CARE OF YOUR DOG	85
6 LIVESTOCK GUARD DOG BREEDS	97
7 LIVESTOCK GUARD LLAMAS	162
8 TAKING CARE OF YOUR LLAMA	177
9 LIVESTOCK GUARD DONKEYS	187
10 TAKING CARE OF YOUR DONKEY	201
Appendix and Resources	212
Index	219

FOREWORD

Livestock Guardians deals with a topic that is steadily increasing in importance and complexity for livestock producers: How can producers keep their stock safe without damaging the environment? The days of wholesale predator eradication are long gone, and few would argue for the return of those days. The new era demands the coexistence of livestock producers with predators, and to the extent that this coexistence can be made peaceful, both the producers and the predators come out ahead. This book is a detailed look into approaches that can make this paradox a reality.

Jan Dohner has presented the details of a very complex subject by breaking them down piece by piece so that readers can understand all of the small components of a very complicated whole. Predator identification and behavior are detailed here, so that producers can appreciate the need to tailor their strategies for reducing or eliminating losses.

The topic of predation and livestock losses is extremely complex, and that means that the strategies for success are also complex. Jan Dohner does not shy away from the many details, but delves into them and teases apart aspects of both the problems and the solutions that provide for strategies that will contribute to producer success in managing predators to avoid livestock losses.

The safeguarding of livestock is a topic that few Americans had to consider several decades ago, but is one that nearly all livestock producers need to consider today. Strategies for dealing with livestock guarding have evolved and matured incredibly over the last three decades. Advances in facility design and use have developed, as well as insights into the behavior and management of guardian animals. Experience with the use of livestock guardians has matured and developed a sophistication that was unheard of thirty years ago.

Livestock producers now have many choices when it comes to selecting strategies for avoiding losses from predators. Among the most complex is working with livestock guardian dogs. Jan's experience with and study of these dogs has resulted in a detailed look of how they function, paving the way to producer success rather than frustration and failure. Her appreciation of the need for pure breeds that vary in their average behavior and style of work is a great aid to those needing to select a guardian for a specific task.

Guardian animal selection, whether dog, llama, or donkey, is one issue. Managing that animal after bringing it home is also critical to success, and Jan has provided practical steps for each alternative that will contribute to producer success with their chosen guardian.

Few things in life are more frustrating than a livestock guardian that fails in its task. At the same time, few things are more satisfying than watching a really good livestock guardian at work. Jan has provided a wonderfully complete resource so that more and more people can achieve that satisfaction.

D. Phillip Sponenberg, DVM, PhD
Professor of Pathology and Genetics
Department of Biomedical Sciences
Virginia-Maryland Regional College of Veterinary Medicine
Virginia Tech

PREFACE

The use of livestock guard animals has grown tremendously in the last 20 years. Out of desperation in their battles against predators, livestock raisers were willing to try any of the guard animals — llamas, donkeys, and dogs. Unfortunately, here in North America, we initially lacked most of the skills and knowledge we needed to select and train these animals. The failure rate was quite high, especially for livestock guard dogs.

In the beginning, most of us struggled alone to solve the inevitable problems. There were many misconceptions and half-truths about how these animals worked as guardians. As information about livestock guard dogs trickled out of Europe and Central Asia, we tried to adapt the methods we read about to our very different situations. Researchers developed projects and surveys to study guard animals. Through trial and error, and after much observation, we are now arriving at some reliable methods for introducing and managing guard animals. Breeders are selecting animals for this specific job, which further contributes to the success rate. People who successfully use these guard animals are communicating with each other and sharing techniques and tips. There is now solid information that can help new users succeed, and there is assistance for any individual problems and situations that will occur.

You can be successful using guard animals if you take the time to learn how these animals work. Provide a good supportive environment that encourages their positive behaviors. Guard animals are not a magic cure for predator problems. The use of guard animals requires an investment in good fencing. You may also need to employ other methods of discouraging predators from your pastures and barns, especially during times of heavy predator pressure.

Don't commit to using a guard animal unless you will enjoy having him on your farm. Take the time needed to master new skills in handling and caring for your guard animal. Animal husbandry is particularly rewarding when you allow yourself to fully enjoy your working relationship with your animals. A guard animal can live a long time, and the investment in learning how to care for him and building your relationship will give you many years of security on your farm.

1
THE PROBLEM OF PREDATION

The relationship between predator and prey is one of the basic elements of nature, but understanding and appreciating this reality of life in a nature film is one thing; discovering the evidence in your pasture or chicken house is another. Since humankind first began to tend domestic animals, we have had to protect our animals from predators. Curiously, it was a predator, the dog, that was our earliest domestic partner.

As sheep and goats stepped into the domestication dance with humans, dogs had already established their role as working animals, and they were soon enlisted in the continual struggle against predators.

Today, predators are a far greater threat to livestock and poultry in the United States and Canada than most people realize. Unless you raise these animals yourself, you are probably unaware of the damage and death predators cause among flocks of sheep and goats, herds of cattle, and yards of poultry. Although methods of predator management such as good barrier fencing and night confinement can help protect livestock, they are often too expensive or too impractical in the areas of the country where most livestock is actually raised. On the range areas of Texas and other western states, pastures are very large and sparse, which means that grazing animals are stocked at low numbers per acre on very large tracts of land. On smaller farms elsewhere in the country, most owners have other jobs that take them away from their animals for many hours every day. In some areas of the country, the large predators have been restored; wolves, bears, and mountain lions can be a formidable challenge to stock raisers. Finally, everywhere in the country most farmers now need to be concerned about coyotes and free-roaming dogs.

In fact, the monetary losses caused by predation are often one of the main reasons producers of sheep or goats decide to abandon their efforts to raise livestock. In the western states, in range situations where producers may use no predator management techniques, lamb losses can be as high as 29 percent, while kid losses can be as high as 64 percent, according to the America Sheep Industry Association. Adult animals are not safe from predation either, with losses as high as 8 percent in adult sheep and 64 percent in adult goats in areas without predator management.

In the United States at least 250,000 sheep and 150,000 goats are lost each year due to documented predation. The actual number of losses is much larger, since most people do not report the loss of

Evaluating Predator Risk

Predation studies often employ the statistics compiled by the National Agricultural Statistics Service (NASS), which conducts regular surveys of more than twenty thousand producers, including both large and small operations. The producers who respond to these surveys tend to use some type of predator management, and they therefore show lower rates of loss than the livestock industry as a whole. Nevertheless, the losses can still be quite dramatic when an individual farmer or rancher applies them to his or her own situation; for example, a 10 percent loss due to coyotes would mean the death of one sheep every year in a small hobby flock of ten animals. Every livestock raiser should consult the NASS predator loss statistics to discover the potential threats in his or her area.

an animal unless they believe that they are eligible for compensation from the government. Some states actually compensate a claim for a single verified predator death at triple the value of the single animal, assuming that to be the more accurate loss. In many cases the loss of a single animal poses a mystery, as the animal simply disappears or the owner cannot identify which predator attacked the animal. The available statistics, however, give us a picture of the relative threats.

Coyotes, with a range that is rapidly expanding, account for approximately 60 percent of total predator losses. Many people are surprised to learn that free-running dogs are responsible for another 13 percent, followed closely by wildcats — mountain lions (also known as cougars or pumas) and lynx or bobcats. Other predators include bears, foxes, wolves, feral hogs, and raptors (such as eagles). Smaller predators generally cause more problems for poultry raisers; weasels, opossums, raccoons, minks, otters, and even skunks prey on domestic birds.

Obviously, where you live determines which predators you are likely to encounter, although all predators are increasing in both number and range throughout the United States and Canada. Even the large predators, once hunted to near extinction, are making a comeback under the protection of the Endangered Species Act. Meanwhile the coyote population, originally found only in the western plains, has exploded across the United States and Canada, both because this species filled the niche left empty in the food chain when large predators disappeared and because coyotes have found ways to adapt to suburban and even urban environments.

It is important that you become familiar with the predators in your area. Ask your neighbors to tell you about any predators they see in the area, and keep an eye out for local news stories about predator sightings. Also check the online information sites of your state department of natural resources, local agriculture services, and livestock producers' groups.

Losses of Sheep and Goats to Specific Predators in the United States

Predator	Sheep and Goat Losses	% of Total
Coyotes	135,600	60.5
Dogs	29,800	13.3
Mountain lions	12,700	5.7
Bobcats	11,100	4.9
Bears	8,500	3.8
Eagles	6,300	2.8
Foxes	4,200	1.9
All other predators	16,000	7.1
Totals	224,200	100

Figures from National Agricultural Statistics Service (NASS)

Current popular opinion frowns on widespread hunting or poisoning of predators, leaning instead toward finding ways to live with predators in our environment. For livestock producers this means that predator problems will continue and even increase in the coming years, necessitating an integrated and evolving plan for discouraging predators from your property and your animals.

Signs of Predation

It is important to regularly check on and observe your animals, keeping an eye out for illness or injury. This is good husbandry practice in any case, and an ill or injured animal may attract predators, so you want to attend to problems right away. Another advantage of frequent checks is that without them, scavengers will eventually feed on any carcass, making it nearly impossible to determine the cause of death. When you find a dead animal on your property, the first task is to determine whether it died from natural causes or predation. You also need to look for clues to differentiate between a predator's attack and feeding pattern and the natural scavenging of a down or dead animal.

When you find a carcass, run through these questions:

- Are there signs of a struggle, such as torn wool, hair, or feathers, spots of blood, drag marks, or damaged vegetation?
- How are the remaining animals behaving? Are they more vocal, nervous, or scattered, or not following usual behavior patterns?
- Are there any bite marks on the animal? Where are they, and what size are they? You may need to clip wool or hair away from the neck to look for puncture wounds. Bites made to a live animal will show bruising or hemorrhaging under the skin.
- Is there a lot of blood? Profuse bleeding occurs only before death and for a short time afterward. Animals that die from natural causes do not show bleeding but might show a loss of bodily fluids, such as urine.
- If the animal is a newborn, do soft membranes cover its hooves? If these membranes are still

ON THE FARM

Judy Steffel
Barking Dog Goat Farm, Houghton, Michigan

Judy Steffel lives way up north in the Upper Peninsula of Michigan, where the winters are snowy and long. A breeder of KyiApsos before she moved up north, she took several adult dogs with her and successfully introduced them to goats. The pups she breeds now go out with the goats when they are a few weeks old.

The 14 dogs at Barking Dog Goat Farm have serious predators to face, including coyotes, wolves, black bears, and mountain lions. They always work in groups, with some sleeping in the barn and others positioning themselves outside at night. For people who need several dogs, Judy advises, "*Watch* the dogs. Think about why they do what they do. Remember that each dog is an individual, a little different from all the others. And observe the pack structure. Although it's tempting to treat them all the same in order to satisfy a human sense of fairness, that feeling does not compute to them."

Judy's personal challenge has been learning to live with a pack of primitive dogs. As she observes, "These dogs are so very close to their Tibetan roots that their instincts are intact. I was fairly knowledgeable about dogs before I met these guys, but the learning curve has been tremendous. They do things that puzzle me — some I've had to ponder for *years* before figuring out!"

Judy's advice to a new owner? "Worry less about what breed of livestock guard dog to get and more about finding a good breeder. In addition to making every attempt to produce healthy, functional dogs, a good breeder provides 'tech support' — for years and years, if necessary — primarily for the good of the puppies she's bred, but also to the advantage of the new owners of those pups. New owners should ask lots and lots of questions and be willing to stick it out through the long puppyhood of the livestock guard dog!"

present, the newborn was probably stillborn, rather than killed by a predator. What shape are the lungs in? The lungs of a stillborn animal are dark in color and will not float in water, while pink lungs indicate that the animal was breathing before death. Is there milk in its stomach? If so, the newborn was able to nurse before death. A field autopsy can answer these questions.

Large Predators

Large predators include coyotes, free-roaming dogs, wolves, mountain lions, lynx, bobcats, bears, and feral hogs. If you have only recently moved to an area where these large predators live, you may be apprehensive about dealing with them. In most cases you are unlikely to actually see or confront them. However, learning about their habits and signs of their attacks on other animals will help you analyze any situations that do arise. In the event of an actual confrontation, please exercise extreme caution for your own personal safety, especially if you interrupt an attack or inadvertently corner or threaten a wild animal.

COYOTES

When European colonists first arrived in North America, they did not see a single coyote *(Canis latrans)*. Until the twentieth century, the coyote lived only in the grasslands and prairies of the western portion of the country. Lewis and Clark first spotted the coyote in South Dakota in 1804 and gave

---

Health Threats from Wild Animals

Wild animals can carry a number of serious diseases that affect both domestic stock and human beings. Any warm-blooded animal, including all canids, pigs, raccoons, opossums, skunks, and their relatives, can potentially carry rabies. About seven thousand cases of rabies are discovered in animals each year in the United States. Ninety percent of these cases are found in wildlife. Raccoons account for about 45 percent of these cases, followed by skunks with 30 percent. Signs of an ill animal include unusual viciousness, confusion, lethargy, aimless wandering, or, in a nocturnal animal, being out during the day.

Feral pigs, whose numbers are increasing, carry a number of serious diseases including pseudorabies (a viral disease), swine brucellosis, tuberculosis, anthrax, tularemia, and trichinosis, as well as a wide range of parasites. If you have feral hogs in your area, find out which diseases your local authorities have identified as concerns.

Some diseases and parasites of wildlife pose particular threats to working animals. Coyotes, for example, can suffer canine health problems including parvo, distemper, mange, and hookworms, all of which can be transmitted to your dogs. If you have horses on your property, you should discourage opossums from your pastures, barns, and feed storage areas. The feces of the opossum (and to a lesser extent the skunk and the armadillo) carry a protozoan called *Sarcocystis neurona*. When horses ingest this protozoan, whether through grazing or eating contaminated food, they can develop equine protozoal myeloencephalitis, a devastating neurological disorder.

Be extremely cautious when approaching any wild animal that appears sick or when handling dead animals. Always wear rubber or plastic gloves; dispose of them properly afterward, or wash and disinfect them, as well as any tools you may have used. Bury carcasses deeply so that they cannot be dug up. Do not feed them to any animals.

COYOTE RANGE
Although the original range of the coyote was confined to the grasslands of the western continent, it is now found throughout all parts of the United States, both rural and urban, as well as south into Central America and north into Alaska and Canada.

it the name *prairie wolf*. They shot one and sent its bones and hide to President Jefferson so that it could be scientifically studied. Eventually the campaign to eradicate the true wolf in the East left a distinct void in the ecosystem for a predator, and the coyote moved right in. By the 1970s coyotes were reported to be living in Kentucky and Indiana, Rhode Island and West Virginia, as far north as Nova Scotia, and as far south as northwestern Florida.

The coyote has now expanded its range south into Central America and north into Alaska. It has adapted to every sort of habitat, including urban, and we have no firm estimate of its total population. Attempts to eradicate coyotes are generally ineffective, with the most successful efforts being aimed at individual problem animals rather than wholesale reduction of the population.

Coyotes can appear yellowish red to gray in color. Coyotes at higher elevations are generally darker gray, while desert-dwelling coyotes are tawny or dull yellow with some gray and brown. Northern coyotes have longer coats. Coyotes have slender legs, a long pointed nose, large pointed ears, and a full bushy tail. They measure 3 to 4 feet long from nose to tail. Curiously, the Eastern coyote is larger than its Western relatives. Some biologists believe this resulted from interbreeding with wolves in Ontario. Generally, coyotes weigh between 20 and 30 pounds, but some Eastern coyotes can be as large as 45 to 60 pounds.

Coyotes eat prey from the size of insects to large ruminants. Fifty percent of coyotes' diet consists of rabbits and rodents, and another quarter comes from scavenged carrion. They also enjoy fruit and can be a bane to watermelon farmers. When they are limited to small prey animals, their social units tend to consist of just two or three animals. With regular access to larger prey, the pack or family size can increase up to ten animals on their home range or territory. Occasionally coyotes live as solitary animals, leading a more nomadic life while in search of a mate or a territory.

Hunting Habits of Coyotes

Coyotes are opportunistic hunters and are the most common predator of farm animals, from poultry to calves. They also often kill domestic cats and small dogs. They tend to attack smaller livestock on the neck, head, or back, resulting in considerable hemorrhaging and bite marks in these areas. They usually attempt to kill a larger animal by attacking the throat, relying on strangulation or shock to bring the animal down. However, younger inexperienced coyotes may bite an animal in other places, more in the manner of a dog attack. Larger animals may also be attacked in this way when a group of coyotes is involved in the kill.

Coyotes begin feeding first at the abdomen behind the ribs, consuming the internal organs, followed by the meaty areas. Coyotes will also chew bones. They will return to scavenge the carcass for several days if given the opportunity. If the carcass is small, however, they may take it away or consume it entirely, leaving behind little evidence of the attack. Coyotes generally kill only one animal at a time unless they are training their young to hunt.

Coyotes usually hunt from late afternoon until dawn. They can climb or jump many types of fences but often prefer to scratch out an area under a fence and slide underneath. They tend to kill animals near the edges of pastures, where they can find better cover.

Lambs and kids are easier prey for coyotes than adult sheep and goats. They do not usually take cows and calves, but if they do, it is typically during or just after birth, when the cow is weak, or during the next few days, when the cow leaves the calf to find food or drink. Coyotes often kill calves by biting through the anus into the abdomen. Calves missing their tails, known as bobtail calves, are a sign of an attempted coyote kill. Where they have access to them, coyotes also will take piglets of up to about 30 pounds.

DOMESTIC DOGS

When domestic dogs attack livestock or poultry, the owners of the guilty parties almost never believe their sweet family pets could cause such damage. And in most cases the dogs were chasing animals for fun rather than predation. Nonetheless, domestic dogs cause enormous damage and injury to livestock and poultry, second only to coyotes.

Despite the popular image of roaming packs of feral dogs, most attacks are carried out by friendly family pets that are allowed to run at large. According to a large Australian study in 1996, 51 percent of dog attacks are carried out by a pair of dogs, usually a male and female from the same home or neighboring homes, while another 40 percent are by a single dog. Dogs of all ages, from three-month-old puppies to elderly canines, of all sizes, spayed, neutered, and intact, purebred and mixed breed, have been caught chasing and killing livestock.

Hunting Habits of Dogs

Dog attacks have many characteristics that distinguish them from attacks by wild predators. For example, the attacks can occur at any time of the day or night. Also, dogs will attack animals of any age and condition, rather than focusing on the small or weak. Dogs chase the animals for fun, often running them into exhaustion or cornering them against buildings or fences. Instead of leaving a single carcass, dogs will injure or kill a number of animals. They tend to bite the flanks or the hindquarters, although they will also bite legs, heads, ears, and other body parts indiscriminately, so that the animals appear mutilated. The animals that are attacked may have broken legs or blood on their hind legs; if they die, they do so from shock or blood loss, rather than a single traumatic injury. Dogs often do not consume any of the carcasses, and the dead or injured animals will be scattered around the pasture or poultry yard rather than dragged away. If dogs do feed on

Dog Attack: What to Do

If you witness a dog attack, write down an accurate description of the dogs, noting the characteristics of the ears and tails, the coat color and length, any markings, the dogs' sizes, and so on. Pictures are invaluable; take photos of the dogs if possible, and definitely take photos of the carcasses. If the dogs are friendly and allow themselves to be caught, restrain them, and call the animal authorities rather than the owners. If the dogs run away, try to follow them calmly. They will probably walk away unless they are chased. Most dogs eventually return home, which is generally located downwind of the livestock. Do not confront or warn the dogs' owners, who will generally deny their dogs' involvement or create an alibi for their dogs before the authorities arrive. Determine in advance which authorities to contact in your area — local animal control or a policing agency.

In many areas farmers have the right to kill dogs that attack domestic animals; however, these laws differ in various states, provinces, and municipalities. Please remember that if you shoot someone's pet, you may be involved in a legal action. Discourage dogs from wandering onto your property by maintaining your fencing and using guardian animals. Do not encourage neighbor dogs to visit your property by feeding or playing with them. Diplomatically ask your neighbors not to bring their dogs onto your property when they visit. It is preferable to have this conversation with neighbors up front rather than having an ugly incident destroy your friendly relationship. Most importantly, always keep your own family dogs from chasing stock.

---

a carcass, they may chew on various parts rather than feeding in the abdominal cavity first as a wild predator would.

For all of these reasons, a single dog can cause far more damage to livestock than a single coyote kill. Farmers whose animals are victims of dog attacks often find these scenes far more emotionally upsetting than kills by wild predators. One or two dogs can do such horrible damage to so many animals that it is hard not to imagine that an entire pack was responsible for the damage.

There are other clues that may be found at the scene of the attack. After a dog attack, the flock or herd will appear far more traumatized and anxious than after a single kill by another type of predator. In the case of sheep, tufts of wool will be torn loose and scattered around the area. Droppings can also be telling: dog droppings are usually brown in color and smooth, rather than showing fur and bones like those of a wild predator. And dog tracks differ slightly from coyote, fox, and wolf tracks. Dog tracks are round in shape and usually as wide as they are long, while coyote tracks tend to be longer than they are wide. A dog leaves marks with all four claws, while coyote tracks show marks from only the front two claws.

Dog and coyote tracks differ only slightly. Dog tracks (left) are usually flatter and rounder, with four deep claw marks. In general, coyote tracks (right) are tighter, more oval in shape; often only the front two claws leave impressions.

Attacks by truly feral dogs are difficult to distinguish from domestic dog attacks; however, there are some useful observations. Feral dogs tend to be wary of people, although they may be very aggressive if they have been shot at or chased by humans. They are active only from dusk to dawn, rather than at all times of day. They often travel and live in packs. And unlike domestic dogs they will consume their kill, generally eating the hindquarters and internal organs.

WOLVES

Before European settlement of this continent, gray wolves *(Canis lupus)* ranged throughout much of the United States and Canada wherever there was sufficient prey. Estimates of their population at that time range from one to two million. The colonists brought with them a historic, powerful, and deep-rooted fear of wolves, one that probably never equated to their actual threat to people. In fact, most wolves are reclusive and avoid human confrontation. Nevertheless, over the centuries, loss of habitat and eradication efforts drove the wolf from most of its native territories. The most intensive efforts at wolf eradication, funded by the government, occurred in the nineteenth century when humans moved into the Great Plains, although wolves were also hunted for their valuable fur.

By the time environmentalists' interest in native species gained strength in the 1960s, there were no gray wolves left in the continental United States outside of extreme northeastern Minnesota and Isle Royale in Lake Superior. Biologists and naturalists have worked over the past few decades to restore the wolf to some of its former range, an effort that has met with a great deal of controversy. As a result of reintroduction programs, gray wolf populations in 2005 had increased to about 3,000 in Minnesota, 400 in Wisconsin, 400 in Michigan, 90 in northwestern Montana, 300 in the greater Yellowstone area, and 500 in Idaho, as well as very small numbers in Arizona and New Mexico. Canada is home to 50,000 to 60,000 wolves, and about 5,000 wolves are found in Alaska. Although gray wolf recovery

GRAY WOLF RANGE
Once ranging throughout much of North America, gray wolves are still present in Alaska and much of Canada and have been restored to the upper Great Lakes region, the greater Yellowstone area into Idaho and Montana, and Arizona and New Mexico.

has been a conservation success, it has increased contact between livestock and wolves and remains a subject of heated debate among ranchers, environmentalists, and other concerned parties.

North America is home to several subspecies of the gray wolf, including the Mexican gray wolf, the Great Plains wolf, the Rocky Mountain wolf, the Eastern timber wolf, and the Arctic wolf. (There is some scientific debate over possible separate species status for the Eastern timber wolf, which is found in the United States and Canada.) Color and size varies with locale. Male wolves generally range from 80 to 120 pounds and are 5 to 6 feet in length; they tend to be smaller in the upper Midwest. The gray wolf is often gray, but it can also be black, brown, reddish brown, mixed gray (grizzly), or white.

Hunting Habits of Wolves

Gray wolves live in packs usually consisting of four to eight animals but including as many as thirty in some parts of Canada and Alaska. Pack territories range from 25 to 1,000 square miles, depending on the availability of prey and the presence of other wolf packs. Wolves prefer large prey — primarily deer, elk, moose, and bison — although they also hunt smaller animals such as rabbits. In Minnesota the average wolf consumes 15 to 20 adult deer in a year, meaning that a pack of six may kill 90 to 120 deer annually. While a single wolf generally brings down a large animal by biting its throat, a pack of wolves works together, biting the prey's nose, anus, legs, and belly. Once the animal is down, the wolves tear at the abdomen to reach the internal organs and feed upon the hindquarters. They will eat smaller animals entirely, either at once or taking away to feed them to pups, while with a larger carcass they will return to feed again.

Biologists do not fully understand why some wolf packs ignore livestock and others become habituated to killing domestic animals. Because it is difficult to keep them from preying on livestock once they have begun to do so, problem wolves sometimes need to be removed by legal authorities. Livestock producers in wolf territory can also employ both preventive and nonlethal techniques to help reduce predation.

Red Wolves

Red wolves *(Canis rufus)* once roamed the eastern forests of North America, but by 1970 fewer than one hundred individuals survived in the coastal areas of Louisiana and Texas. As pressure from hunting and habitat loss increased and their numbers continued

---

Keeping Wolves Away

Very large predators such as wolves and bears can be too much for one livestock guard dog to handle alone. Most people who raise livestock in large-predator territory use at least two and sometimes three or more dogs, along with good fencing. Other than grizzly bears and mountain lions, wolves present the greatest danger to working dogs. If you anticipate wolves moving into your area, it is better to establish your own "pack" of working livestock guard dogs before that occurs.

In general, reports reveal that wolves tend to avoid areas that are already marked as territory by other canines. If you have an existing wolf problem, you will likely need more dogs to discourage the wolves from the area, especially if they already regard it as their territory. In addition, your fencing may need to be reinforced by two or more strands of electrified or barbed high-tensile wire outside of your existing fence.

to fall, red wolves started breeding with coyotes, and scientists began an effort to capture the remaining pure red wolves to preserve genetic purity. Though captive breeding programs have allowed breeding populations to be reestablished in North Carolina and on some barrier islands in the Atlantic, there is continuing controversy about the genetic purity of these animals. The total free-ranging population is still only about one hundred animals, with a somewhat larger number in captivity.

Red wolves are 3 to 4 feet long and weigh 40 to 80 pounds. They are often brown in color with buff

Coydogs, Wolf-dogs, Coywolves?

Wolves, coyotes, jackals, and domestic dogs all have 78 chromosomes arranged in 39 pairs, which makes hybridization or crossbreeding among them possible. Although wild canids and domestic dogs have existed side by side for thousands of years in North America, there is little actual evidence of widespread interbreeding. There are several obstacles in the way of such interactions. For one thing, wolves and coyotes are more likely to kill domestic dogs than mate with them, and dogs are generally aggressive toward coyotes and wolves.

For another, wolves live in a pack structure, and coyotes are often secretive or solitary, coming together only to mate and raise pups; neither situation is amenable to the inclusion of a dog. And successful puppy rearing for both wolves and coyotes usually demands that the parents have an established relationship, unlike the case for dogs. Finally, female wolves and coyotes come into heat only once a year, approximately January through March, and the males are fertile only during those times as well. Female dogs, on the other hand, come into nonseasonal heat usually twice a year, which explains why male dogs are capable of breeding year-round.

So any conceivable scenario would require an unusually receptive female coyote or wolf and a nonaggressive male dog meeting at just the right time. Coydogs and wolf-dogs may exist, but in far smaller numbers than popular culture suggests. When supposed coydogs are examined, most prove to be unusually colored coyotes or feral dogs.

Curiously, the questionable penchant for purposefully bred wolf-dogs sold as pets may contribute to situations in which these crossbreeds escape or are abandoned and become feral. And where do these wolf-dogs come from? Virtually all the wolves used in these breeding programs have been captives for several generations, despite the claims of some breeders. Generally a male dog is bred to a female wolf. As to the tales of Inuit sled-dog breeders using wolves to strengthen their dogs, it's simply not likely, since sled dogs have been bred for a desire to pull through many generations of selective breeding. The Alaskan malamute and Siberian husky were not developed by interbreeding with wolves in any recent history.

Coyotes and wolves can interbreed, and have, as evidenced by the recent red wolf–coyote crosses and the possible Eastern coyote–wolf breeding in the past, which may have contributed to the larger size of the Eastern coyote. This is far more understandable given their common biology.

Wolf-dogs and coydogs are not a good idea as family pets and are a terrible idea as farm dogs. Livestock guard dog breeds have been selectively bred for centuries for an inhibited prey drive, and for obvious reasons a wolf-dog or coydog would definitely not make a good livestock guardian.

HISTORICAL RANGE OF THE RED WOLF
Once occupying a much larger range, red wolves are now exceedingly rare. The population may be limited to as few as a hundred animals.

or red and black shading. They can be confused with coyotes at first glance, although red wolves have a more powerful build and a larger footprint than coyotes.

Red wolf packs tend to be small, as are their territories. The animals are nocturnal and often hunt alone or in pairs. They hunt deer, raccoons, rabbits, and rodents, and eat eggs, insects, and some plant food. At this time, the red wolf is not an important threat to livestock.

MOUNTAIN LIONS

The solitary, tawny mountain lion *(Puma concolor)* goes by many names: *cougar, panther, puma,* and the old-fashioned *catamount*. Historically the mountain lion ranged throughout the Americas, from southern Alaska south through Central and South America. By the beginning of the twentieth century, however, the mountain lion had been exterminated throughout eastern and midwestern North America, except in Florida, and driven into unpopulated or extremely rugged areas of the West. Today, the number of mountain lions in the American West is rebounding, and the animals are also migrating back into the Midwest, with occasional sightings in the East. The mountain lion population in the United States is estimated at about 30,000; these animals are now also found in good numbers throughout many areas of central and western Canada. Mountain lions can roam a home territory from 25 to several hundred square miles.

Mountain lions are substantial in size, with females weighing 80 to 130 pounds and males up to 180 pounds. They have long, heavy tails. Fencing will not deter mountain lions, as they can make a vertical jump of up to 15 feet and a horizontal leap as long as 40 feet. The only absolute protection against mountain lions is to house livestock and poultry securely at night in buildings or covered enclosures.

Hunting Habits of Mountain Lions

These large cats are solitary hunters that most often prey on deer, which they take down from behind.

MOUNTAIN LION RANGE
The mountain lion once roamed throughout the continent. Today it is found in the western portion of its former area but it is migrating eastward in both Canada and the United States. Isolated sightings are increasingly reported in the midwestern states.

An adult typically kills a deer once every week or two. Mountain lions also take other prey, including birds, rodents, smaller predators, and, unfortunately, domestic pets, sheep, goats, and larger livestock such as cattle and horses. Biologists report that young, inexperienced mountain lions, especially males, are much more likely than older lions to prey on livestock.

Mountain lions hunt from dusk to dawn. They attack larger livestock from the rear, biting the back of the neck to bring the animal down and then suffocating it. Mountain lions may bite smaller livestock on the head, sometimes breaking the neck. A single mountain lion will sometimes kill several smaller animals in one night. Animals killed by mountain lions will have tooth punctures about two inches apart and claw marks on or around the neck; large bones may be broken. A mountain lion will feed on the front or hindquarters and leave the abdomen alone. It may drag the carcass of a larger animal quite a distance from the kill site and return to feed on it over several days, covering it with grass, leaves, dirt, or other debris between visits. Cats have retractable claws, unlike dogs, coyotes, and wolves, so mountain lion footprints generally lack claw marks unless the ground is very soft.

LYNX AND BOBCATS

The Canadian lynx *(Felis lynx)* is found throughout Canada, western Montana, and neighboring areas of Idaho and Washington, and may also be present in Colorado, Oregon, and Wyoming. Small numbers are found in New England.

The lynx has tufted ears and long hair with yellow-brown to frosted gray coloring, with occasional dark spots. Its tail is shorter than that of the bobcat and has a prominent black tip. Its large furry pads help it move on snow. The lynx can be as tall as 24 inches and can weigh up to 35 pounds. Although it can be about the same size as a large bobcat, the lynx generally looks bigger due its longer legs.

The typically smaller bobcat *(Felis rufus)* ranges from southern Canada through Mexico, although it is most common in the southeastern and western

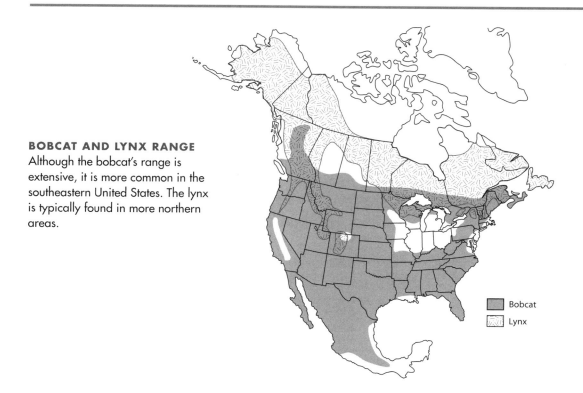

BOBCAT AND LYNX RANGE
Although the bobcat's range is extensive, it is more common in the southeastern United States. The lynx is typically found in more northern areas.

United States. The bobcat's fur is shorter than that of the lynx and colored buff to brown, with darker stripes and spots. It has tufted black ears, a facial ruff, and a short tail. It stands up to 21 inches tall and weighs 13 to 30 pounds. Female bobcats are markedly smaller than males. Size seems to increase with higher elevations and northern latitudes. Larger bobcats can be difficult to distinguish from Canadian lynx. If you are attempting to distinguish a small bobcat from a feral or domestic cat, note that a bobcat appears two or three times larger than a cat, and its hind legs are longer in proportion to its body than a cat's.

Hunting Habits of Lynx and Bobcats

Nocturnal creatures, lynx generally prey upon rabbits, rodents, and similar-size animals, though they do also attack deer and caribou calves. They kill deer and other large animals by leaping on the back and biting the neck; these attacks leave claw marks as well as bite marks. The prey dies from suffocation or shock. Lynx typically attempt to cover the carcass for later consumption. They will also feed on deer carcasses left by hunters or carcasses of domestic animals. The secretive lynx is less likely to come near areas of human habitation than the bobcat, although there have been some cases of lynx attacks upon sheep and goats.

Bobcats prey primarily upon small animals such as rabbits and birds, but they can take down a deer by jumping on its back and biting the neck. Hunting mainly during dusk or dawn, bobcats will attack smaller livestock, and they are capable of climbing into roosts to attack domestic turkeys. Bobcats bite the head, neck, or throat on smaller animals and leave claw marks on the body. They seem to prefer the hindquarter flesh and internal organs, often leaving the rumen uneaten. They may also drag away their kill and attempt to cover it.

BEARS

Black bears *(Ursus americanus)* are found throughout most of North America. Though they are typically found wherever there is thick vegetation in

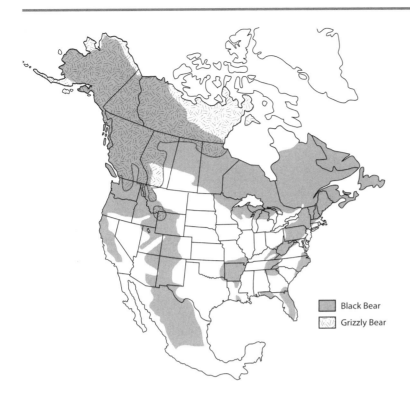

GRIZZLY AND BLACK BEAR RANGE While the grizzly bear remains a protected species in the United States and parts of Canada, black bears are becoming increasingly common in eastern areas.

mountainous areas, their range is expanding, and they are now present in small numbers even in many suburban areas. Black bears are indeed usually black in color, with a light-colored muzzle and sometimes a white spot on the chest. However, in the West they are often lighter in color, ranging from brown to reddish brown to blonde. Black bears in coastal areas of Alaska and British Columbia are cream or bluish gray in color. The average size of a male black bear is 250 pounds, although individual bears can be considerably larger. Females average about 150 pounds. Eastern black bears are larger than their Western counterparts.

The more aggressive and less predictable brown bear *(Ursus arctos)*, including its subspecies the grizzly bear *(Ursus arctos horribilus)*, is found in Alaska and in the Rocky Mountains of both Canada and the United States. Although it can be difficult to distinguish them from black bears, brown bears and grizzly bears have larger shoulder humps, shorter ears, and a concave or dished profile. Grizzly bears, which live in the Rocky Mountains, have on their shoulders and backs longer hair with lighter-colored tips, which gives their fur a frosted appearance. Brown bears vary in size according to habitat and food sources. Adult males can be 300 to 900 pounds, while females are quite a bit smaller at 200 to 450 pounds.

Bear tracks are distinctively flat-footed with five toe impressions. Claws will show only in soft mud. The hind foot mark somewhat resembles a human foot.

Hunting Habits of Bears

When bears emerge from hibernation in spring, food is scarce and they are hungry for protein and fat. They may attack livestock at this time. They kill by breaking the back or neck, leaving large claw marks on the neck or shoulders. They may kill several animals at a time. A distinctive sign of a bear attack is that bears skin their prey. They do not eat the rumen or stomach, and they tend to leave the hide and the skeleton intact rather than scattering them around or eating the bones. They may move their kill to a more hidden place before consumption. Bears leave clearly identifiable footprints and scat.

ON THE FARM

Peter and Joan Sanders
Houseman Acres, 100 Mile House, British Columbia

Peter and Joan Sanders live on 240 acres deep in the interior of British Columbia. Coyotes and bears are their primary predators, although mountain lions are found in the area as well. The Sanderses utilize several techniques to protect their stock. They have good, tall fencing, which can be quite a challenge to maintain in this heavy snow country. Peter creates safe paths through the snow for the sheep to use, which seems to keep them from wandering near the perimeter fencing. He also blows the snow from the area on either side of the paddock fencing to preserve the fence's integrity during deeper snows.

A key precaution is bringing the sheep in close to feed at night, training them through regular grain feedings. Heated water troughs also help keep them close to the shelters in cold weather. During the winter, the ewes and lambs are confined in this area. Newborn lambs and their mothers are also kept inside the barn in individual pens for a week or so, until they are less vulnerable to predation. Another good fence surrounds these feeding paddocks and shelters. The area is well lit at night to discourage the coyotes.

The Sanderses leave their rams with the ewes until just before birthing season, as they believe that the more aggressive rams provide additional protection against predators. The rams spend the winter in another enclosure area outside of the ewes and lambs, providing another layer of protection for the lambs.

Despite the fact that the Sanderses regularly see both coyotes and bears on their property, their gelded male llama, Vinnie, has done an excellent job protecting their sheep for the past four years. The Sanderses have observed Vinnie chasing solitary coyotes and even bears from the summer pastures in daylight. Vinnie lives with the ewes year-round, except when the lambs are very tiny, since he wants to play with them and he is just a little too boisterous. The Sanderses have learned to recognize his alarm call and alert behaviors when predators are nearby. The farm dogs will also sound an alarm.

The Sanderses believe that livestock raisers need to share information about local predators with their neighbors, as a somewhat informal "predator watch." In addition, they try to get to know people with similar stock, since they are a source of useful information. Peter relates, "It works both ways. We hold an informal open house a couple of times a year for our neighbors. We exchange stories and tips, and it is a pleasant social gathering."

Black bears are primarily vegetarians that feed at dusk or dawn, although they can become habituated to human-created food sources and alter their feeding times. When their preferred foods are scarce, they may attack livestock. Although they are capable of killing horses or cattle, they tend toward smaller livestock such as sheep, goats, and pigs. Black bears are generally less of a threat to livestock than brown bears.

Brown bears are incredibly strong; they can kill large animals with a single blow, run faster than a horse, and drag large carcasses a long way, even uphill. They are omnivorous, with grizzly bears tending to be more carnivorous. Brown bears can kill prey as large as a deer; grizzly bears, in particular, are capable of killing animals as large as moose and elk. When brown bears attack larger animals such as moose or bison, they generally choose calves or weakened individuals. They tend to feed more actively at dawn and dusk, although grizzly bears can be active during the day as well. In some areas of Idaho, Montana, and Wyoming, grizzly bear predation has severely affected the raising of livestock.

Livestock raisers can discourage bears with large mowed and cleared buffer zones around pastures and other animal areas. Electric fencing and noisemaking devices are also useful as negative conditioning. Livestock guard dogs will charge bears, barking furiously and discouraging them from crossing fence lines. There is anecdotal information that brown bears are more discouraged by the presence of livestock guard dogs than grizzly bears are.

FERAL HOGS

Four million feral hogs *(Sus scrofa)* currently range across thirty-nine states, and wildlife ecologists predict that their range will continue to expand, as wild hogs have no natural predators and are opportunistic feeders and ready breeders. Generally they are the descendants of domestic pigs that simply escaped, although in the past it was common practice for farmers to let their hogs run wild most of the year, rounding them up with dogs at slaughter time. In some areas, such as New Hampshire, North

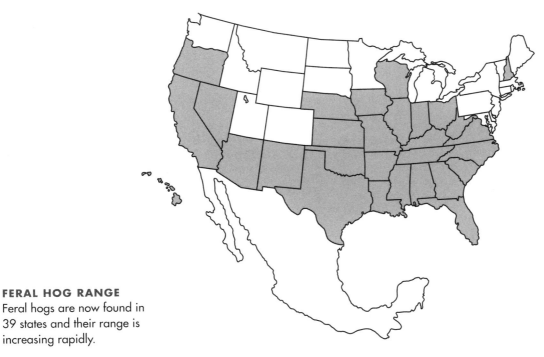

FERAL HOG RANGE
Feral hogs are now found in 39 states and their range is increasing rapidly.

Carolina, Tennessee, and Texas, true wild Russian boars were imported and released as game animals. They still roam wild today, and in some places they have interbred with escaped domestic pigs. Certain counties in Texas and neighboring Oklahoma have the greatest problem with feral hogs, hosting a population of one to two million animals. California, Florida, Hawaii, Mississippi, North Carolina, Tennessee, and West Virginia all designate feral hogs as legal game animals, and in most areas feral hogs are considered an exotic (nonnative) animal and are not protected by law.

Feral hogs may grow as tall as 36 inches at the shoulder, weigh 100 to 400 pounds, and have four tusks up to 5 inches in length. Color varies widely due to their domestic roots; they may be all one color, spotted, or even banded. Russian boars are usually light brown or black with tan underparts, longer snouts and legs, and shorter, straight tails.

Hogs are active from dusk through the night but not during daylight hours. Their tracks resemble those of sheep, goats, and deer, which can add to the difficulty of identifying them. Some signs that hogs are in your neighborhood include large areas of rooted soil, mud wallows, and rub marks on trees. Hogs create other concerns besides livestock predation for farmers, since they can destroy crops, pastures, and fencing.

Hunting Habits of Feral Hogs

Hogs are true omnivores, and they have an excellent sense of smell. While not a major threat to livestock in most areas, feral hogs do prey on weakened or ill cattle, goats, and sheep, as well as newborn animals. They are especially attracted to livestock birthing areas, where they will consume stillborns, afterbirth, and other fetal tissues. Hogs are much more likely to prey upon twin lambs than upon single lambs, since they are smaller and their mothers have more difficulty watching over multiples. Hogs will also chase flocks and attack newborns by rushing them, knocking them over, and trampling them.

A hog will bite or crush the skull to kill the animal. The hog starts to consume the newborn in the belly area, eating the stomach or rumen, sometimes after skinning the carcass. Given time, a hog will usually consume the entire carcass, leaving no evidence other than a damp or bloody area. In this way hog predation can sometimes contribute to the illusion of low herd production. Some experts believe that a significant number of livestock deaths blamed on coyotes are actually due to hog predation, especially in Texas.

Small Predators

The smaller predators found throughout most of North America can pose a danger to lambs, kids, and poultry. Surprisingly, raptors and other large birds are more likely than foxes to kill or harass newborn livestock. The small carnivores and omnivores — foxes, raccoons, opossums, weasels, minks, skunks, fishers, feral cats, and rats — are more likely to prey on poultry than on other livestock. Free-range poultry are likely victims for raptors and large birds, while the small mammalian predators will invade poultry housing.

EAGLES, HAWKS, AND OWLS

Owls are night predators, while hawks and eagles are daytime hunters. Free-range poultry is a prime target for these birds, but young lambs and kids are also vulnerable. The golden eagle, the largest predatory bird in North America, with a wingspan of up to 2.5 meters, can be a threat during lambing and kidding season, particularly in the western states. Attacks on calves occur much more rarely. Raptors will also feed upon carrion, however, so it can be a mistake to automatically assume that the birds you find eating a carcass actually killed it.

Hunting Habits of Raptors

All of these raptors attack from the air by deeply puncturing the prey's head with their talons, which leaves distinctive marks there, but the rest of the body intact. Raptor talon punctures will be deeper than mammalian tooth punctures and may reveal the pattern of three front toes and one hind toe.

Owls often remove or eat the head of poultry. Hawks and eagles may leave deep puncture wounds on the head, neck, back, and breast. Raptors usually pluck birds before eating them, which can help you determine whether they actually killed or merely scavenged poultry: the base of a plucked feather will be smooth or clean if it was plucked soon after death, while small amounts of flesh usually cling to the base of the feather if it was pulled off a cold carcass. Owls and other raptors can carry off smaller poultry, leaving no evidence of the attack.

OTHER LARGE BIRDS

Ravens, vultures, crows, magpies, and gulls usually scavenge carcasses rather than kill, although they will attack newborns, weak animals, or birthing animals. They will also consume afterbirths. Turkey vultures are entirely carrion eaters, but black vultures can occasionally be predatory. Black vultures have a wingspan of up to 1.5 meters. Their population is increasing, as is their range, which now extends up from the southern and southwestern states into the Midwest, New York, and Connecticut. They live in communal roosts and often feed together in groups. Black vultures will attack young calves, sheep, kids, pigs, and even cows giving birth.

Hunting Habits of Large Birds

These birds locate their prey by sight. They begin their attacks by pecking at the eyes or head, often blinding the animal, puncturing the head, or fracturing the skull. They also attack the nasal, navel, and anal areas. Magpies, in particular, have been known to create wounds on the backs of sheep. Lacking talons (or, in the case of vultures, having only weak talons), these birds do not carry off their prey.

> ### Raptors Are Protected
>
> **State and federal permits** are required to trap or shoot owls and hawks; these permits are given only after nonlethal methods have failed and there is a serious depredation problem. Similarly, it is illegal to kill a vulture without a permit from the U.S. Fish and Wildlife Service. Golden and bald eagles are strongly protected under federal law, although the Fish and Wildlife Service occasionally gives trapping permits for specific cases where golden eagles are preying on livestock.

FOXES

The red fox *(Vulpes vulpes)* is found through much of the world, including Australia, where the British introduced it. It has adapted to a wide range of habitats, including suburban areas. The red fox is by far the most common fox in North America, and it is more likely than the other true foxes to prey on livestock and poultry. Although a member of the Canidae family like the coyote, wolf, and dog, the red fox cannot interbreed with these other canids, because it has only 38 chromosomes, rather than the 78 the others possess.

At first glance red foxes look like slender, lightly built dogs weighing only 6 to 15 pounds. They vary in color from light yellowish red to darker reddish brown, or even black, with lighter-colored underparts. The lower legs are generally black in color, and the tail is tipped in either black or white. Red foxes do not form packs but lead solitary lives.

They eat rodents, rabbits, and fruit and will also scavenge carcasses. Red foxes can be active during the day but are more likely to approach human habitation during the night. They are great leapers, able to clear a fence as high as 6 feet.

Hunting Habits of Foxes

Foxes are most likely to attack poultry, including turkeys. They often carry birds away with them, leaving behind very little evidence of the kill. They usually eat the breast and leg meat first. They also eat eggs, doing so quite neatly and leaving the shell nearly intact.

Foxes may prey on newborn lambs, though their small size requires them to kill by multiple bites to the throat. They tend to eat the internal organs, feeding through the belly. They do not cause severe damage to the bones, as their jaws lack the crushing power of their larger canid relatives.

MISCELLANEOUS SMALL PREDATORS

In many areas of the country the following small predators are commonly found around farmsteads. They are common and persistent predators of poultry. If you do not house your poultry securely, you are sure to lose birds to these predators every year. Some of these predators will return night after night to vulnerable poultry until you have lost most of your birds. If you use livestock guard dogs to protect your farmstead, you will find carcasses of raccoons and opossums outside your barns or in your fields, since livestock guard dogs will readily kill these two small predators.

Raccoons *(Procyon lotor)* are nocturnal omnivores that eat a large variety of small animals and insects, as well as fruits and vegetables, including those found in your garden. On farms they prey primarily on poultry and to a much lesser extent on lambs or kids. They can kill newborn lambs by chewing on their noses. Raccoons are often attracted to caged birds and other small animals. They sometimes pull the heads of birds through mesh or wire fencing and bite them off, either eating the heads right there or carrying them away. In this manner they can kill several birds in one night, especially if they are feeding their kits. If they capture a whole bird, they may eat only the head and crop, although they have been known to eat entire birds. They also will take eggs, usually carrying them away to be eaten elsewhere.

Although not a major factor in livestock loss, raccoons are the most common wildlife source of rabies, which they can transmit to domestic animals. For this reason many horse owners are now inoculating their animals against rabies, although such practice may be cost prohibitive for other livestock. The best defense is to discourage raccoons from your buildings and pastures. Exercise extreme caution around raccoons that have been trapped or that act sick and also around raccoon carcasses.

Weasels, minks, and skunks, all members of the Mustelidae family, are found in most areas of North America and are a threat to poultry. Weasels are the most destructive members of this family. They attack poultry on the sides of the head or the base of the skull. They will kill several birds at once, often leaving the bodies after they consume the head and neck, although they may eat the internal organs. Sometimes they will try to drag dead birds away, but they often leave the carcasses in a pile.

Minks kill in a manner similar to weasels but more often prey on waterfowl, since minks live on riverbanks and the shores of lakes. In spite of their small size they are capable of killing much larger poultry, including ducks and geese.

Skunks sometimes prey upon hatchlings, but they particularly relish eggs, eating the insides after making a hole in the shell. They do not carry eggs far from the nest to eat them, and an egg that has been eaten by a skunk can appear as if it has hatched instead. Skunks, like raccoons, can be a source of rabies and because of this can be a threat to larger livestock.

Fishers *(Martes pennanti)* are also members of the Mustelidae family. Like weasels, they will bite off the heads of poultry, but they are capable of taking animals far larger, including lambs, kids, and even domestic pets. Although fishers' primary range lies in Canada, they are present in the Northwest down to California and can be found in the Northeast as well.

Opossums *(Didelphis virginiana)* are nocturnal omnivores that prey on poultry, both birds and their eggs. They usually attack only one bird, and they eat the bird through the abdomen, consuming the stomach. They may eat a young or small bird entirely. When opossums eat eggs they crush them, leaving only small pieces of shells in the area.

Domestic and feral cats will prey on birds and rabbits. Feral cats are extremely shy and typically hunt at night, but domestic cats will hunt anytime. Domestic cats may not eat what they catch, while feral cats hunt to feed themselves. Cats may consume very small animals entirely. When consuming larger birds, cats eat the meaty parts, leaving the skin and feathers. Waterfowl, both domestic and wild, are the most vulnerable to cats.

Rats will readily prey on very young poultry and occasionally on adult birds and newborn livestock. Their burrows, tunnels, and droppings will be visible in the areas they frequent. Rats can transmit numerous bacterial and viral diseases and should be controlled on every farm. They also damage grain storage areas and electric wiring in barns, which probably poses a larger threat than predation to livestock.

Historically, *predator control* meant the lethal removal of predators. In recent times, however, lethal methods have become less acceptable to the public. Methods include shooting, hunting with dogs, trapping, snaring, aerial hunting, and the use of baited cyanide injectors (mechanical devices known as M-44s that inject cyanide into the predator's mouth, resulting in near-instant death) or livestock protection collars (rubber bladders containing sodium fluoroacetate, which kills a coyote within five hours after ingestion), and den hunting to kill pups (commonly known as denning). Poison bait, once widely used, has been illegal since 1972. These lethal predator control methods are sometimes unavoidable, however, especially in the case of an ill or wounded predator or one that has turned unusually vicious or is impossible to discourage.

Lethal control methods are not effective unless they are highly selective, which can be very difficult to accomplish. Simply undertaking a campaign to exterminate local predators in general is unlikely to be successful. In the case of coyotes, our most common predator, resorting only to lethal means of control would be a never-ending battle. It is estimated that we would need to kill 60 percent of the coyote population each year before we would see any reduction in coyote population density. In areas such as the Edwards Plateau of Texas, where coyote predation is high and ranchers consistently have needed to control them vigorously, coyotes have become extremely wary and nearly impossible to trap. Coyotes are now so prevalent throughout North America that whenever they are removed, new coyotes move right in. Since coyote mortality is already high (in national parks, annual coyote mortality is about 50 percent), often any coyotes that are hunted or trapped would have died anyway.

Wildlife biologists recommend killing or capturing only confirmed problem predators, leaving nonproblem animals undisturbed. Experienced farmers advise against killing coyotes that are not preying on your livestock, because their replacements may already be experienced livestock predators. Also, if the alpha male or female of a pack is killed, the young will become desperate and will certainly prey on your livestock. Removing the breeding pair from a pack will only increase the number of new packs, as the young adults will be forced to fend for themselves.

Nonlethal Predator Control

Truly nonlethal methods of predator control include secure housing and fencing, good husbandry and animal management, and various predator-frightening techniques. Livestock guard animals can provide primarily nonlethal protection, but at times, especially in the case of livestock guard dogs, the protectors will kill some predators. A combination of these methods is the most successful and most effective way to increase your level of protection. A well-considered, integrated plan of protection will significantly reduce predation and will often prevent it, depending on your location, livestock, and predators.

Current research has concentrated on alternative methods of predator management and control such as disruptive stimuli (page 33) and aversive stimuli (page 36). Because coyotes are the primary predator of domestic livestock, most efforts have been directed toward coyote control, but with the increasing numbers of bear, wolf, and mountain

Check Your Local Laws

Many of the lethal predator control methods are illegal in or regulated by certain states, provinces, and municipalities. In addition, laws concerning wildlife vary by municipality. Before undertaking the extermination of any problem predators, make sure you know the local laws that apply to the control methods you intend to use and the species you will use them on. And please remember that predation by protected or endangered species is a matter for the legal authorities, despite the real danger these animals may pose to livestock producers.

lion contacts, efforts are now being made in those areas as well. Undoubtedly, as time goes by many of these research efforts will provide more concrete and practical suggestions.

Some livestock producers prefer nonlethal means of predator control because they believe in sustainable methods of agriculture or they enjoy living with wildlife, while still discouraging undue losses among their own animals. These producers value the role predators perform in controlling the populations of rabbits, rodents, and deer that damage crops. Other producers want to be sure that their neighbors and the public see them in a positive, nonthreatening way. Others choose sustainable methods of predator control because of the research that demonstrates the futility of lethal methods. Most will pick and choose their way through all the available methods of predator control, tailoring their choices to fit the needs of their own farm.

Farmers and ranchers who try to practice sustainable agriculture attempt to work with nature, acknowledging the interconnection between farming and the environment. They often accept a small amount of predator loss, even as they strive to prevent it. One benefit of using sustainable methods is that some consumers will pay more for the end

A combination of nonlethal predator control efforts is more effective than any one method. An integrated plan of protection will reduce predation.

products because they believe those methods are both ethically and environmentally conscious.

The social concepts of animal welfare are also important to both these producers and their consumers. As these beliefs become more widely held, predator-friendly labeling is becoming a desirable niche market. The effort to produce a "Predator Friendly" certification process and marketing label began in 1991. A group of ranchers, clothing manufacturers, and conservationists began to certify woolgrowers who pledged not to kill native predators on their land and to use both livestock guard animals and other pasture management strategies to minimize confrontations between stock and predators. In 2003 this group turned over the certification process to the Predator Conservation Alliance. (For more information on predator-friendly marketing and how to achieve this labeling tool, see the appendix.)

Nonlethal Methods for Preventing Losses of Sheep and Lambs

Nonlethal Method	Percentage of Users (in Total Reporting Group*)
Fencing	52.5
Guard dogs	31.8
Guard llamas	14.0
Guard donkeys	9.1
Lambing sheds	30.8
Active herding	5.7
Night penning	32.9
Fright tactics	2.2
Removing carrion	11.7
Frequent checks	14.0

*More than one method is frequently used.
National Agricultural Statistics Service (NASS)

Practicing Good Husbandry

There are many different practices and techniques that you can utilize in protecting your animals from predators. None is completely reliable. However, the more methods you adopt and practice, the greater the likelihood you will be successful. Choosing a variety of methods allows you to select what will work best on your land, with your animals, your production needs, and the predators you need to confront. You may need to utilize the more intensive practices only during short periods of predator stress, such as livestock birthing season or times when predators are having difficulty finding their natural prey. Here are measures you can take to make your property less attractive to predators:

- Remove all rubbish, brush piles, and vegetation near animal housing, pens, and fence lines.
- Keep your pastures and fencerows mowed, since more attacks occur where there is tall or rough cover.
- If you are raising poultry, locate their housing and yards away from trees and other high perches that can be used by raptors.
- Secure animal feed so that it does not attract wild animals.
- Fence in gardens and orchards.
- Use lighting to make your property seem less safe to predators.
- Do not feed or encourage deer in your area.
- Check your animals regularly, looking for ill or injured individuals that might attract predators.
- Isolate sick or ailing animals in a safe location.
- Make sure that household garbage and animal waste such as carcasses and afterbirths are disposed of so that they do not attract wild animals.

The importance of removing all dead livestock and wildlife cannot be overemphasized, although it is admittedly difficult or impossible on large ranges. Groups of small producers in an area will often work together to keep a large area free of carrion.

It may sound simplistic, but you should count your animals frequently. If you do not, predator losses could occur over several days without your knowledge, reducing your ability to prevent more losses. Livestock raisers need to remind themselves that herding was once an active daily job. Many animals were let out to graze only in the company of herders. Active herding is still practiced among the large range flocks where the pressure of predators is very heavy.

Record keeping is also important. If you suffer a predator attack, record the date and location. Pastures located near water often have more coyote attacks than other pastures. More losses are recorded in larger pastures than smaller ones. Hilly or rough pastures provide more cover for all predators, as does brush and tall grass, and most predator kills occur in the roughest areas of the pasture.

If you are using a rotational grazing scheme, add the pressure of predators into your management plan. Monitor your pastures to discover which ones are more vulnerable at certain times of the year, and make sure you are aware of when specific predator-attracting activities such as birthing are occurring. Take stock of the factors in your area that affect predator pressure, such as when prey is in short supply for predators and when predators have young to feed or train to hunt. Some producers choose to change the dates of their lambing or calving seasons so that they don't coincide with times when predators are particularly active. This tactic is especially useful for cattle raisers, since calves are vulnerable to coyotes for only a short time.

The use of sheds for lambing or kidding and secure lots for calving close to the farm center significantly reduces the losses of newborns and mothers to predators. Owners of small- to medium-size flocks commonly move new mothers and their babies into individual pens inside a barn or a shed for a few days until the initial high danger from predators has passed. The penning also facilitates close monitoring for potential medical problems and bonding between mothers and offspring. The use of confinement is certainly more labor intensive for a short time, but the results are generally worth the trouble. If your animals are not able to lamb, kid, or calve in a protected area, choose your birthing pastures carefully, thinking about potential predator problems just as you would consider protection from weather and increased chance of disease.

Night penning also reduces predator loss, especially when coyotes or wolves threaten sheep and goats. All livestock can be conditioned to expect nighttime feeding and trained to enter corrals or paddocks willingly. The level of security around night pens can often be higher than in pastures, with the use of more substantial fencing, electric wires, lighting, and guard or farm dogs in the vicinity. With the animals closer to your home, you will be far more likely to hear any disturbances as well. The one caution about night penning is that it concentrates the animals in one location where dogs or even coyotes may do considerable damage in a single attack.

PROTECTING POULTRY

Poultry naturally roost in the evening and can be easily cooped at night. Train your birds to return to their coop at night by keeping them confined full-time inside the coop for several weeks while they are growing up. Even when they range outdoors during the day, keep their feeder and waterer inside the coop.

Making the coop secure presents challenges. If you are constructing a coop from scratch, install concrete floors with one or two rows of concrete block footings to prevent rodents, snakes, and other small predators from digging under the floor and walls. Seal all openings greater than one inch in size to prevent minks and weasels from entering the coop.

Where you need screening, utilize heavy mesh wire screening, which is more secure than chicken wire. (Remember that raccoons can reach through traditional chicken wire to grab poultry by the throat, and foxes can enter through net or wire mesh greater than three inches in size.) Amazingly, raccoons can open many simple latches, so take care in devising fastenings for windows, doors, and other openings.

Poultry runs should be situated away from trees and tall brush, which provide cover for predators. Runs also need to be securely fenced. Wire fencing buried one foot deep, with an underground apron bent outward six inches, will prevent most predators from digging under the fence. It may be necessary to use an electric wire, placed four to six inches off the ground, to discourage persistent predators. Netting or mesh wire over the top of a run provides protection from raptors and climbing predators; another possibility is an overhang of fencing, electric wire, or barbed wire around the top to discourage climbers.

If you use chicken tractors or other mobile housing for your poultry, place them inside well-fenced pastures that are patrolled by livestock guard dogs. Free-range turkeys are usually not cooped at night, so fencing for these birds needs to be especially well built and secure.

If you have mountain lions in your area, the recommended fencing is chain link with a canopy cover or roof. Bears can be discouraged by sturdy electric fencing at least three to six feet high, with maximum protection afforded by a fence that is eight or nine feet.

Fencing Guidelines

Fencing is your first line of defense for all livestock and poultry. Good fencing is essential for containing livestock guard animals, especially dogs. There are two types of fences: *exclusion* and *drift*. The purpose of an exclusion fence is to keep wildlife *out*. Exclusion fences are designed differently depending on which animals they are meant to keep out. For example, deer exclusion fences need to be very high, while coyote or fox exclusion fences need to have narrow openings to prevent these animals from squeezing through them. Drift fences are designed to keep livestock *in*. Again, the design of the fence depends on the type of animal it is meant to confine. A drift fence for cattle, for example, can be as simple as a few strands of barbed wire, but it will not prevent other animals from passing through it.

Some producers construct a double ring of fencing around the pasture, leaving room for livestock guard dogs to patrol the entire perimeter between the two fences. In this situation the outer ring is an exclusion fence and the inner ring is a drift fence.

While classic board or rail fence looks lovely, it provides virtually no protection against predators. Fences that do provide predator protection include wire mesh, electric, and electric net fencing. Barrier or boundary fencing is generally electric, wire mesh, or a combination of these materials. Before you make your fencing choices, ask yourself these questions:

- Is there existing fencing on the property, or do you need to construct new fencing?
- If there is existing fencing, can it be reasonably upgraded?
- Which predators are present in your area?
- What is the replacement value of your animals?
- What is the lifetime value (as compared to the one-time cost) of your fencing options?
- How rugged, rocky, dry, or wet is the land you need to fence?
- Are you constructing barrier fencing or fencing around a night pen or paddock?
- Is the area too large for expensive fencing choices?
- Do you need to consider the needs of migratory animals?

FENCING OPTIONS

Fences made of barbed wire or non-electrified high-tensile wire will keep many livestock animals in their pastures, especially cattle and horses, but they will not keep predators out. In most cases, it is easier to keep your animals inside a fence than to keep predators out. There are several types of fencing that you can use, depending on the kind of animals you raise and the geography of your area.

Wire Mesh Fencing

Wire mesh fencing should be considered for boundary or perimeter fences since it remains effective if the electricity fails. Mesh fencing also provides a physical barrier to animals, both your stock and wildlife. As an example, both deer and horses will *see* a mesh fence, and they are much less likely to run into it by accident or in a panic than they would a few strands of wire.

A 5½- to 6-foot mesh fence with a mesh overhang and apron will serve as a basic exclusion fence. Mesh deer fencing, with its greater height of 6½ to 8 feet, may be needed in some areas.

An existing wire mesh fence can be upgraded to make it more secure. One or more strands of electric or barb wire added above the existing fence will increase its height and prevent climbing by dogs and coyotes.

Wire Spacing for Electric Fences

Animals or Purpose	Number of Wires	Wire Spacing (in Inches)*	Post Spacing (in Feet)**
Drift Fence			
Cattle with calves	3	13 - 11 - **12**	90, 65, 50
Sheep with lambs	5	**6** - 7 - **8** - 9 - **10**	30, 30, 25
Goats	6	**5** - **5** - 6 - 7 - 8 - **9**	50, 50, 35
Poultry	3	7 - 7 - 7	30, 30, 25
Exclusion Fence			
Raccoons and other small animals	3	**4** - **4** - 5	90, 65, 50
Bears	4	**4** - 5 - **6** - 20	20, 20, 15
Foxes and feral pigs	6	2 - **6** - 6 - **8** - 8 - **9**	30, 30, 25
Maximum Security Fence			
Overall predator control	8	4 - **5** - 5 - **5** - 6 - 6 - 7 - **8**	30, 30, 25
All livestock and most predators	10	4 - **4** - 4 - **4** - 5 - **5** - 5 - **5** - 5 - **6**	50, 50, 35

*__Bold numbers__ indicate the heights at which wires should be electrified.
**For level, rolling, or steep terrain

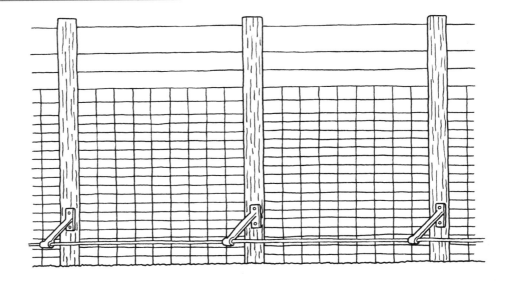

Good predator exclusion fencing is made from wire mesh woven more tightly near the bottom, with additional electric wires above the wire mesh to discourage jumping or climbing and an electric trip wire outside the fence to discourage digging.

Mesh fencing should be strung tight to the ground and on the inside of the fence posts. Although more costly, the newer type of high-tensile wire mesh fencing reduces sag and allows for greater distance between posts. Traditional mesh fencing should last 15 to 20 years; high tensile fence should last 30 years.

Electric Fencing

Electric fencing is generally less expensive than mesh fencing and more adaptable to rough or rugged land. An electric fence works as an incomplete or open electrical circuit. The energy source sends electric pulses through the wires. When an animal comes in contact with a charged wire, its body grounds the fence, allowing electricity to travel through its body to the soil and on to the nearest grounding rod, and from there back to the energy source to complete the circuit. The animal experiences the passage of electricity through its body as a shock. Vegetation or other objects touching the charged wire can also create a closed circuit, which results in reduced voltage in the fence and little or no shock to an animal that comes in contact with it, so fence lines must be kept clear.

More intense shocks are delivered from a system that uses alternating charged and uncharged, grounded wires. When an animal touches both a charged and a grounded wire, the electricity travels through the animal and directly back to the energizer through the grounded wire. The grounded wire is a much better conductor than soil, so the animal receives a stronger shock.

Modern electric fence chargers or energizers can be operated by battery, solar cell, or the domestic power supply (which may or may not be the most reliable source in some areas). The newer low-impedance energizers are not as affected by tall weeds touching the fence wires, dry soils, or snow, all of which contribute to voltage and current loss, as they respond to these conditions by producing higher amounts of energy, although wet vegetation or woody brush will still decrease the line voltage. The low-impedance energizers also reduce the risk of fire, as compared to the older "weed burner" electric fence chargers.

Electric fencing requires a grounding system and lightning protectors. Line voltage should be checked regularly with a voltmeter. Corners and long spans need to be well braced to support the tension required for these fences. Wire tighteners will facilitate the maintenance of proper tension. And all electric fences should be clearly identifiable as such by anyone who might walk up to them.

Less expensive than wire mesh, electric fences should be designed appropriately for both your stock and your predators.

To discourage bears, wolves, coyotes, and dogs — the most serious predators — perimeter electric fences must have at least five wires (but keep in mind that wolves and coyotes have been known to jump electric fences shorter than six feet). The lower three wires should be six inches apart, and charged and grounded wires should alternate. As you increase the number of wires, you increase the effectiveness of predator protection. A nine-wire fence is very predator proof. Twelve- or thirteen-wire exclusion fences, five and a half feet tall, are considered extremely predator proof, although quite costly.

Energizers should last at least fifteen years, while the high-tensile wire should last thirty years. In addition to wire, electric fencing can be constructed from tape, braid, or rope, although these may not have the same electrical resistance or life span. On the other hand, some of these alternatives are much more visible than plain wire.

Electric Net Fencing

Electrified net fencing is not usually considered permanent, nor is it commonly used for perimeter fencing. It is most useful for subdividing pastures or otherwise limiting animals' range on a temporary basis, such as in rotational grazing, for cleaning up crop fields, or on leased land. Electrified poultry net fencing works well for free-range grazing, as it protects poultry against coyotes, dogs, bobcats, and raccoons and other small predators. Netting does pose a risk to both livestock and wildlife, due to the danger of entanglement. All animals kept in netting should be trained to respect it. The cost of netting is higher than that of conventional fencing materials, and its life span can be quite short, but it is extremely easy to move.

INSTALLING A NEW FENCE

As you plan your new fence, try to keep fence lines as straight as possible, and clear all trees and brush from the fencerow. Note that it is very helpful to be able to drive a vehicle along either side of a fence, for mowing and other purposes.

Try to avoid crossing rough areas such as waterways, ravines, and unstable or soft ground. If you do need to cope with these troublesome areas, make sure you use appropriate measures to cross them. Since coyotes, dogs, and other predators will attempt to squeeze through gaps at the bottom of fencing, it is important that you prevent large gaps from occurring. If you are using wire mesh fence, you may need to construct a skirt that conforms to the shape of

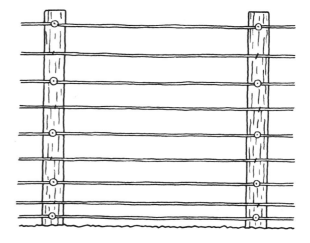

A properly designed and constructed nine-wire electric fence provides good predator protection.

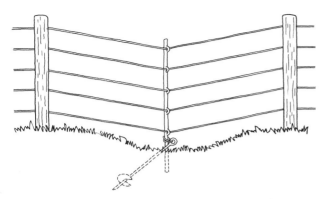

An extra post can anchor wire through a shallow dip, minimizing gaps.

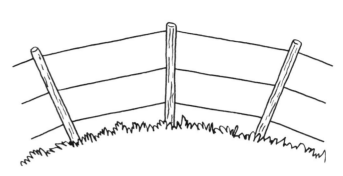

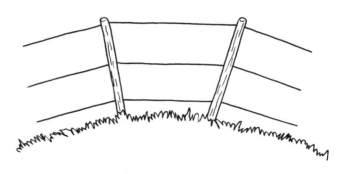

A properly constructed fence over a rise maintains adequate fence height (top). Due to improper post placement, the example below doesn't maintain good fence height.

the gap and attaches under the fence. A skirt can be hammered together from treated (exterior-grade) wood, or a frame of wood or steel can be filled in with wire mesh, high-tensile wire, or another sturdy material. You may need to anchor the sides or bottom of the skirt to the ground. You can also run a scare wire outside the fence about eight inches off the ground, or even electrify an entire skirt, to discourage predators from approaching the barrier.

If you need to allow for occasional rushing water through a ravine or gulley your fence runs through, you may need to construct a free-swinging gate at the site. To keep animals away from this gate, you can install a scare wire or electrify the entire gate. Alternatively, you can construct the gate as an electrified curtain of 9-gauge galvanized chain, with the suspended lengths of chain spaced six inches apart. To prevent an individual gate from draining power from the entire fence, you can insulate each gate and power it separately through a water gate controller. A floodgate isolator or energy limiter will shut off the current temporarily during rising water. These devices are available from electric fence suppliers. You could also install a manual cutoff switch for use during times of flooding or rushing water. After any serious flooding or high water, you must check these gates for damage and to clear away any debris.

UPGRADING AN EXISTING FENCE

If you have existing mesh wire fencing in reasonably good shape, you can probably upgrade it economically to make it more secure against predators. Good mesh fencing has wires spaced horizontally six inches apart and vertically two to four inches apart. It may have a tighter mesh at the bottom to exclude smaller predators and contain smaller livestock. It should be five to five and half feet tall to discourage climbing or jumping by coyotes, dogs, lynx, and bobcats (although coyotes have been known to climb or jump six-foot fences). If the

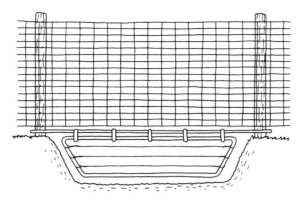

A free-swinging water gate can be constructed with a frame of welded steel or wood, filled in with woven wire or high-tensile wire.

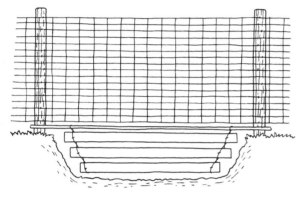

A water gate can also be constructed by stapling galvanized wire to treated boards sized to fit the space. Twist the wires together between the boards to increase stability.

Two strands of wire add height to this mesh fence while the electric trip wire on the outside discourages predators from climbing over.

existing fence is shorter than this, one or two electric wires spaced six inches apart above the mesh can provide the necessary height. Barbed wire can be also be used, but it is less effective and potentially damaging to larger livestock, which may lean over the fence to graze.

Another option for reinforcing old fencing is to place two or three electric wires in front of the mesh, outside of the pasture, attaching the electric wires to the existing fence posts. And of course you should repair all areas of damaged mesh promptly, since large holes will allow predators to crawl through the fence.

To prevent predators from digging under the fence, place barbed wire near ground level or bury a wire apron under the fence. Or, even better, install an electric trip wire located six inches off the ground and eight to ten inches outside of the existing fence. An electric trip wire will also discourage canines from traveling along the fence and deter small predators. Many producers feel that the electric trip wire is the single best investment to improve any fence. As mentioned earlier, vegetation must be kept down around the electric wire in order not to short out the fence.

GATES ARE IMPORTANT

After careful attention to constructing a new fence or improving an existing fence, don't forget the gate! Gates are a common weakness in keeping predators out and livestock in. Traditional board or pipe gates are meant to keep large animals such as cattle and horses in, but they do little to keep predators out. In addition, these heavy materials can be difficult to hang and move. Galvanized rail gates are lighter in weight, which allows you to place the rails closer together, helping to keep predators out and smaller livestock inside the enclosure. Wire-filled gates are the best choice for excluding predators and keeping your small livestock and poultry where you want them.

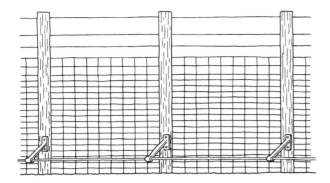

An electric trip or scare wire can be used outside the pasture fence to discourage predators from digging under the fence. Placed inside the pasture, a trip wire will prevent livestock guard dogs from digging under the fence.

When hanging pipe, wood, or metal gates, try to avoid leaving spaces at the sides of or under the gate. You may need to reinforce a wood, pipe, or rail gate with wire mesh over its cross members. Electric wire can be used above a gate to prevent predators from climbing it. Scare wires can also keep animals away from the gate area. Some folks bury a heavy-gauge livestock panel one to two inches deep under the gate area to discourage predators from digging under it.

Disruptive Methods of Predator Control

Disruptive stimuli, which include visual, auditory, and other repellent techniques, are intended to frighten the predator. The value of these techniques is that they provide an immediate response to a predator attack and can be useful during times of increased predator stress, such as calving or lambing season. The problem with these techniques is that the predator eventually becomes habituated to them and then ignores them. However, disruptive stimuli can provide effective short-term protection until a more permanent solution is found. Their usefulness is extended if they are moved or changed frequently.

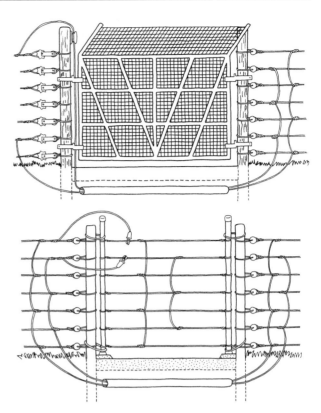

Wire-filled gates (top) are the best choice for excluding predators. Fit the gate tightly to the space. A wire mesh overhang will prevent climbing. In situations where you don't need regular access, a secure exclusion gate can be created from electrified wire (bottom). These drawings also show how to maintain electric current around gates.

USING VISUAL DETERRENTS

Visual techniques can be quite simple: lights left on at night, vehicles moved into pastures, even the age-old scarecrow. Night lighting over corrals or pens is generally effective in discouraging coyotes, and it does not seem to disturb livestock. However, there is slight evidence that it may facilitate dog attacks, which is understandable, since family pets would not be frightened by artificial light. Night lighting also allows the farmer or rancher to check more easily on disturbances and repel predators. There is no specific proof that strobe lights work better than regular lights, although motion-activated lights can be useful, since they appear random to predators. Lights can be placed on timers for convenience. Lighting around poultry housing is also advisable; in fact, night lighting actually increases egg laying by poultry.

Fladry, a line of small flags hung on a rope, is a less well-known visual deterrence. Fladry is used in addition to regular fencing, since it does not provide any barrier to confine stock or exclude predators. Interestingly, fladry originated in eastern Europe as an aid to hunting wolves; the lines of fladry flags form a barrier that wolves will not cross, at least initially. Preliminary research indicates that fladry is useful against wolves for a period of up to two months. After they become habituated to the flags, their usefulness sharply decreases. Fladry would be useful for short-term protection of relatively small, high-risk areas such as birthing grounds. Fladry has not been proved effective against coyotes, bears, or birds.

Fladry flags, originally made from strips of cloth, are now available as plastic or nylon strips,

ON THE FARM

Cherie Faktor
Willow Ridge Farm, Lucas, Iowa

Willow Ridge Farm, on 40 acres of gently rolling hills in south-central Iowa, is home to a number of fiber animals, including Shetland sheep, Angora goats, and even Angora rabbits. There are large packs of coyotes in the area as well as predatory birds. Six years ago Cherie Faktor and her family purchased a male llama and donkeys to serve as livestock guardians. They now have five llamas and four donkeys and recently added a Great Pyrenees livestock guard dog. The llamas are also kept as fiber animals for Cherie's home-based business.

Each guard animal has a particular job. The three jennies watch over the large herd of sheep and lambs, while a jack lives with the rams and stud goats. Two female llamas guard another group of goats. The relative newcomer, the Pyr, stays with the brand-new lambs. Cherie keeps the three jennies together with her sheep because she believes they are more able to handle the packs of coyotes by working together. One of the jennies was born in the sheep pasture and raised with the sheep. Cherie relates that this particular jenny is closely bonded to the sheep and is an outstanding livestock guardian.

The only management problem Cherie and her family have encountered is that the jack is somewhat distracted whenever a jenny is in season. Keeping an intact jack is not recommended for most livestock guardian situations, but the Faktors use him for breeding, and this has worked out in their situation. They are also cautious about using the llamas with their sheep, worrying about potential dangers from a llama attempting to breed a sheep; however, the llamas have been absolutely trustworthy with their goats. Cherie handles her llamas a lot so that they are comfortable and safe with people.

Cherie believes that every animal has a different personality and temperament, so that owners should proceed slowly when introducing livestock guardians to their stock. Sometimes, she says, owners may need to try different animals until they find a perfect fit. Cherie preaches patience: "We know what works with what, and that just comes from watching and learning from them both — the guards and the guarded!"

spaced about 18 inches apart on a line. Although cattle will chew on the plastic strips, the coated nylon strips resist fraying and are less attractive to cattle. Another modern creation is turbo fladry, which is strung on electric fencing material rather than rope and can be electrified; this may extend its effectiveness. Fladry is available at per-mile costs of $500 to $1,200, depending on the material used. Fladry can be made at home by tying foot-long strips of plastic flagging to the top wire of fences. Livestock producers also report some success with homemade fladry items such as aluminum pie plates, balloons, wind chimes, and other objects.

Fladry must be checked frequently in order to untangle any flags that have become wrapped around the rope. Because of the flags' propensity for wrapping around their line, fladry is not practical in high-wind areas.

USING AUDITORY DETERRENTS

Auditory or sound techniques have also proved effective against many predators, including the largest. Sound techniques include radios left on at night, bells on livestock, and horns, sirens, or propane cannons set to go off at random intervals. Talk radio, in particular, has an unpredictable rhythm and works well as predator deterrence when left on all night. Some producers use amplifiers on radios to increase their protective area; others play recordings of distress calls of a particular predator. The use of bells with a herd or flock is an age-old technique that helps herders find their animals at night, and it has been proven that coyotes and bears never attack the belled sheep in a larger flock of sheep without bells.

> ### Combination Devices
>
> **Devices that combine** light and sound are reputed to be more effective than either technique used alone. One such device, known as the Electronic Guard, was developed by the Wildlife Service of the Animal and Plant Health Inspection Service (APHIS) of the U.S. Department of Agriculture (USDA). It is a portable, battery-powered unit that combines a flashing strobe light and siren, both of which activate at random intervals. Research has proven that the Electronic Guard serves as deterrence against coyotes and wolves. APHIS recommends that two units be used in fenced pastures under 10 acres in size and one unit per 10 acres in larger pasture areas. It is effective for one to four months but should be moved frequently.

Fladry provides short-term visual deterrence to predators and is potentially useful during periods of extra pressure.

Gunfire, fireworks, and other noisemakers can be useful temporarily, but they are not practical or perhaps even safe as a long-term solution. And as the sounds you use become louder, they will also become more disruptive to your neighbors and your livestock, not to mention you and your family. It may be necessary to alert not only your neighbors but also law enforcement officials if you plan to use sound devices. Sirens, horns, and propane exploders can be programmed to random intervals to increase their effectiveness. Proponents of propane cannons, which are very loud, advise that the timer be set to explode every eight to ten minutes and that the cannons should be moved every three or four days.

Wildlife will habituate to sound stimuli, so it will become useless over time (although many people believe that talk radio played inside barns or coops continues to be effective over time). Again, save these measures for periods of high predator pressure.

At the other end of the noise spectrum are ultrasonic devices. These have not been proven to be effective at the decibel levels available on the market. The ultrasonic levels necessary to repel predators would also irritate or harm humans, pets, and livestock.

CHEMICAL AND BIOLOGICAL DETERRENTS

Researchers are working to develop chemical and biological repellents for predators, primarily for coyotes, but at this time no such chemical repellents are approved in the United States. One difficulty in developing such repellents is that any material that would be applied directly to livestock also must be safe for human consumption. Producers have observed that approved insecticides, specifically lice medications, temporarily deter predation, but this is not a long-term solution to predation.

One chemical device that is approved for use with livestock is the antipredator collar, which is filled with an irritating or toxic substance, such as capsaicin or sodium fluoroacetate, and placed

This sheep is wearing an antipredator collar filled with an irritating or toxic substance. These collars are not safe for use with livestock guard dogs.

around the necks of small livestock. While the collar can stop an attack that's under way, the animal wearing it is often still injured or even killed, and all members of the flock would need these collars to be protected. You should not use livestock guard dogs with stock wearing antipredator collars, since they could accidentally bite the collar or come in contact with its contents after a predator attack.

Since coyotes and wolves respect scent marking of territories, farm or guard dogs provide a natural biological repellent. Theoretically, artificial scent markings might also deter predators, though the frequency of application poses a problem.

Aversive Methods of Predator Control

Aversive stimuli work by conditioning a predator not to perform a specific behavior. Due to the need to condition each predator and the expense and effort required, these techniques are generally used only on specific individuals of a protected species that are exhibiting problem behavior. One experimental tool is the electronic training collar, similar to the collars used to train dogs. This collar can activate light and sound devices or deliver a shock when the animal wearing it approaches a forbidden area. Unfortunately, predators eventually habituate to the light and sound devices, and the batteries powering the collar eventually expire. At this time, only approved researchers are using aversive collars.

Discouraging predators by offering meat tainted with chemicals that cause illness (conditioned taste aversion) results in animals that are wary of bait but does not discourage the actual killing of stock. Similarly, the use of rubber bullets, pepper-filled paintballs, and other nonlethal ammunition tends to make the predator wary of the person doing the shooting rather than the area being protected. Furthermore, it is illegal to harass some predator species, and doing so can be physically dangerous.

Diversionary feeding is sometimes attempted to condition predators to finding food in a specific location rather than preying on livestock. However, diversionary feeding is illegal in some locations and may actually increase the number of predators.

Mixed Grazing Herds

Cattle can discourage predation on sheep and goats by small predators such as coyotes, particularly in rougher pastures, where sheep and goats are at greater risk. A difficulty with this arrangement is that cattle and sheep or goats do not naturally graze together in a single group. Dean M. Anderson of New Mexico State University, who has done research on combined grazing techniques, has coined the term *flerd* to describe a multispecies group created through pen-bonding lambs or kids to heifers or cows over a period of 30 to 60 days. Pen bonding increases the cattle's tolerance of the smaller livestock, while the smaller animals learn to seek the cattle's presence for protection.

The use of aggressive animals, including donkeys and llamas, as guardians in your pasture is, in fact, a form of multispecies grazing. In addition to the deliberate use of livestock guardians, be alert for those ewes or cows (or other stock) that show an unusual aggressive reaction toward predators. Often

> ### Ineffective Deterrents
>
> There is no proof that electromagnetic devices are helpful in deterring predators. Low-power, long-wavelength laser devices do disperse smaller birds such as gulls, crows, and ravens, but these devices are quite expensive and care must be taken not to aim them at people, cars, or aircraft.

these animals are the least human-oriented, and it is tempting to remove them to make your job easier, but their leadership, instinct, and experience are valuable in your flocks and herds. Consider keeping them, even if they are unproductive, if they are known to chase a coyote across the pasture. If you are able to keep your ram, buck, or bull in your pasture you will naturally benefit from his aggressive and protective instincts, but this may not be possible (for several reasons). Some producers report that a pony or horned cattle such as a Highland can be helpful in deterring predators. In fact, any significantly larger animal pastured with sheep or goats will offer some measure of deterrence.

Making Your Choices

You can see that you have many choices in predator control techniques and tactics. It has been proven that a combination of techniques is more successful than any single technique. The most variable technique, but often the most valuable, is the addition of a guardian animal to protect your stock. A guardian animal is just that: a living animal that will require your care, your attention, and perhaps some careful training. The choice to use a guardian animal should be made seriously and with some forethought and planning. Careful preparation will make the use of such an animal more likely to be successful.

3
LIVESTOCK GUARD DOGS

What are the characteristics of a good livestock guard dog? Researchers have identified three essential behavioral traits: he should be attentive, protective, and trustworthy. A properly trained and socialized guardian stays with his livestock and moves calmly among them. He tolerates humans moving among and caring for his animals. He is friendly with his owners and accepts their handling. He does not bite or chase his stock. He may exhibit nurturing behavior to young animals. He is alert to possible threats and able to evaluate them. He may patrol or mark the area around his stock. He will probably be more active at night than during the day. He alerts the shepherd to potential problems. He attempts to discourage threats through barking and posturing. He exhibits a graduated series of responses and takes aggressive action against a predator only when necessary. He is large and powerful enough to confront a predator. He is often described as an independent thinker.

Livestock guard dogs operate with behaviors developed over centuries of use by humankind to protect domestic flocks. They are not protecting themselves when they attack or chase a predator; they are protecting their territory and/or protecting their stock. They will mark their territory to warn predators away. Some livestock guard dogs, in fact, are very territorial and prefer to protect a particular space; others will readily move with their flock to another pasture.

Livestock guard dogs, which can work individually, in pairs, or even in groups, are your only possible animal defense against wolves, bears, mountain lions, and coyote or wild dog packs. They will kill small predators such as raccoons and opossums, and many dogs will be alert to and bark at large birds flying overhead, which makes them effective protectors of poultry and rabbits as well.

A definition in terminology may be helpful here. A *working livestock guard* is a dog that spends most of his time with stock. He may be on a farm or on the open range. A livestock guard dog may work primarily as a *farm guardian*, interacting with his family rather than spending most his time with stock, but still protecting the farm from predators and threats. Farm guardians usually live outdoors. A livestock guard dog that lives in the house with his family is primarily a *companion* or a house guard.

Historically the livestock guard dog breeds have done all of these jobs, although there are differences among breeds and the cultures in which they developed, so that some breeds may be more or less successful in these roles in the modern world. Most experienced users of livestock guard dogs will tell

A pair of Anatolian Shepherd Dog guardians exhibits calm protectiveness over their sheep.

> ### Pros and Cons: Livestock Guard Dog
>
> **PROS**
> - Can guard a variety of animals, including poultry, sheep, goats, llamas, alpacas, and cattle
> - Protects stock against small and large predators, including the most dangerous
> - Provides graduated response, beginning with barking, followed by charging or posturing, and ending with an attack only if the predator is not driven away
> - Provides long-term protection, as predators do not become accustomed or habituated to dogs
> - Able to analyze threats; is a self-thinker
> - Bonds to stock
> - Alerts owner to threats and disturbances
> - Works with other livestock guard dogs
> - Reduces human labor and allows stock to be out at night to graze
> - Allows use of pastures with predator pressure
> - Protects family and farm
> - Provides predator-friendly control
>
> **CONS**
> - Requires good fencing unless working on open range
> - May wander
> - May harass or injure stock if not properly selected and trained
> - May be aggressive to strangers
> - May be overprotective of stock
> - May not tolerate herding or other farm dogs
> - Requires relatively large purchase price, plus care
> - Requires dog food instead of forage or hay
> - Needs time and guidance to mature properly
> - Will bark, especially at night
> - May dig dens
> - Cannot be used in conjunction with traps, snares, or poison

you that a dog cannot be both a livestock guardian and a house companion. As attractive as it sounds, there are definite conflicts in this concept. If a puppy is raised in the house, he probably will not be happy if he is suddenly taken to a field and left there with stock, without human companionship.

If a puppy is not raised with stock, he may never behave appropriately with them. If a puppy plays with other household dogs or dogs at the park, he may not recognize that stray dogs are not welcome in your pasture. If your livestock guard dog sleeps in your house at night, he may not hear a predator in your field. Certainly there are exceptions to these generalizations, but there is truth to the saying that you can't be in two places at once.

The Origins and History of Guard Dogs

There are 34 species within the family Canidae. Among these species, the dog is most closely related genetically to the gray wolf, followed by the coyote, the golden jackal, and the Ethiopian wolf. Not surprisingly, the dog can interbreed with these closely related species. Most experts actually classify the dog as a subspecies of the wolf, calling it *Canis lupus familiaris*. The gray wolf was originally found throughout the Middle East, Eurasia, and North America. Some experts believe that the dog was probably domesticated from the wolf in eastern Asia, while others postulate multiple independent origins of dogs and possibly frequent interbreeding between early dogs and wolves. As the genetic evidence continues to be unraveled,

Spiked collars first described by ancient Roman writers are still worn by livestock guardians in their native homelands. These Kangal Dogs are in Turkey.

we will learn more about the history of man's best friend and ancient partner.

The dog shares a long history of partnership with man. Studies conflict as to whether domestication occurred 15,000 or as long as 140,000 years ago. In either case, the dog was the first animal to share his future with man. The use of dogs to protect domestic flocks is unquestionably ancient, as shown by depictions in early artwork and writing. Evidence of domesticated dogs has been found, along with signs of sheep and goats, in the areas where the earliest human settlements have been discovered in Turkey, Syria, Iran, and Iraq.

By the era of the Romans, dogs were divided into five kinds: mastiffs, greyhounds, pointers, sheepdogs, and spitz dogs. Roman writers Varro and Columella both describe the sheepdog as being white with a loud bark, and they mention the nail-studded collar he should wear to protect him from wolves. Columella wrote that buying a dog should be "among the first things which a farmer does, because it is the guardian of the farm, its produce, the household and the cattle."

The influence of three of these five ancient types — mastiff or *molosser*, greyhound or gazehound, and spitz — is seen in the livestock guard dog breeds today. The mastiff is a large, powerfully built dog with heavy bone structure, a large head, a strong neck, pendant ears, a blunt muzzle, loose skin and lips, and often a dewlap (a flap of skin that hangs under the neck). This type of dog is seen as early as 3000 bce on Egyptian monuments and 850 bce on Assyrian bas-relief carvings. Equally ancient are the greyhounds, which are built for speed and hunting with long legs, a light and lean body, and a flexible back. The spitz dogs are more lupine or wolflike, with upright, pointed ears, a narrow muzzle, long, thick fur, and a tail that often curls over the back.

These dogs of ancient types, usually called shepherd's dogs or sheepdogs, were used for centuries in places where sheep or goats were a primary agricultural focus and large predators were a serious threat. The sweep of cultures that trained large guard dogs included southern Europe through eastern Europe, the Middle East, and central Asia.

THE EARLY USE OF GUARD DOGS IN NORTH AMERICA

This ancient tradition did not take root in the United States until very recently. In the Northeast, the earliest colonists were British. In Britain the large predators had been primarily exterminated by the fifteenth century. Shepherds were able to turn their sheep loose to graze alone and gather them up when needed with the use of herding dogs, and the tradition of using livestock guard dogs had mostly died out.

The Eastern colonists struggled with wolves throughout the seventeenth century. George Washington raised sheep at Mount Vernon and worked diligently for many years to improve the scrubby colonial sheep type with imported breeds such as the Leicester and Tunis. He often wrote to his friends about the loss of his sheep to both wolves and dogs. More than once he asked General Lafayette to find him some true wolf-hunting dogs from Europe.

Washington also exchanged letters with Thomas Jefferson on this topic, as Jefferson acquired several *chiens de plaine* or shepherds' dogs while in France. Georges-Louis Leclerc, Comte de Buffon, the famed naturalist, described the *chiens de plaine* as the foundation for other types of dogs such as the mastiff and "great hound." Jefferson corresponded with Buffon and was a guest in his home. When Jefferson returned to Virginia, he bred the dogs and gave the puppies away as gifts. In 1809, still in search of good guard dogs, Jefferson wrote to P. S. DuPont Nemours, who was leaving for France, "If you return to us, bring a couple of pair of true-bred shepherd's dogs. You will add a valuable possession to a country now beginning to pay attention to the raising of sheep." Four years later General Lafayette was able to send him more shepherd's dogs, and Jefferson later wrote to him, "The shepherd dogs mentioned in yours of May 20, arrived safely, have been carefully multiplied, and are spreading in this and neighboring states where the increase of our sheep is greatly attended to."

Sheep keeping mostly remained a small enterprise in the New World until the pioneers reached the vast grasslands of the West. The Spanish did bring large shepherd dogs with them into New Spain for use in the mission flocks, but shepherds in the American West never adopted this method of predator control in large numbers and the Spanish shepherd breeds did not survive. As sheep keeping expanded in the nineteenth century, so did the lethal approach to predators.

LIVESTOCK GUARD DOGS IN NORTH AMERICA TODAY

Unfortunately, the big sheep-guarding dogs of the early colonists never gained a foothold in American agriculture and were ultimately absorbed into the great American canine melting pot. In both the East and the West, dealing with the wolf meant killing it. Instead of using livestock guard dogs, shooting, trapping, denning, poisoning, and eventually even aerial hunting of large predators were widely practiced through the 1960s.

In the 1970s, as the public became interested in protecting large predators and eliminating many lethal forms of control, private and government entities began to research the feasibility of utilizing livestock guard dogs to protect stock on farms and ranches. Environmental organizations began

The traditional use of livestock guard dogs never really took root in the New World. Serious interest did not develop until the 1970s.

The Great Pyrenees was the only traditional livestock guardian breed widely available in North America before the 1970s.

to promote nonlethal methods of predator control, and the public adopted an increasingly negative attitude toward lethal controls. The use of poisonous substances also came under greater regulation. At that time the Great Pyrenees was the only traditional livestock guard dog breed present in large numbers in North America. The Komondor and Kuvasz were also present but much less common, and the American Kennel Club (AKC) recognized these three breeds primarily as show and companion dogs. Though there was no written guidance available about using these dogs in their traditional roles, dog-savvy owners began to work with them on their rural homes or farms, and they began to report their successes. Serious research soon began.

Beginning in the 1970s studies of livestock guard dogs were conducted by several researchers, including Ray and Lorna Coppinger of the Livestock Guard Dog Project at Hampshire College, Jeff Green and Roger Woodruff at the USDA-funded U.S. Sheep Experiment Station (USSES) in Idaho, and William Andelt at Colorado State University. The Coppingers imported ten dogs of various types and breeds from Europe and the Middle East and bred or crossbred them, placed them with sheep raisers, and followed up with questionnaires and interviews. The USSES researched the use of livestock guard dogs and observed working dogs. These early experiments were based on anecdotal information or observation and conducted without much real working knowledge of how to train these dogs. Studies are still under way in many areas as producers face growing predation by coyotes and the larger predators.

Earlier adopters struggled with many problems and challenges in learning these new techniques and adapting them to the North American style of ranching and farming. Early on there were several serious misconceptions about these dogs. Owners were told by most early researchers to raise puppies alone with stock, away from all human contact.

If you own a livestock guard dog you are guaranteed to hear this comment: "Oh, a livestock dog. Like a Border Collie."

No, you will patiently explain, a livestock guard dog protects the sheep; he doesn't herd them. Practice your answer — you will need it.

In their native homes, livestock guardian dogs usually watched the flocks in the company of the shepherds.

Carnivore Conservation Groups

There are a number of groups that conduct research and disseminate information concerning predator control with livestock guard dogs and other guardian animals. A great deal of information is available on their Web sites and through their published research. Some of these groups also place livestock guard dogs with livestock producers.

Individuals and groups involved in wildlife conservation have also been involved in promoting nonlethal or sustainable methods of predator control. The USDA, sheep and goat associations, and state wildlife and agricultural agencies promote the use of livestock guard dogs and provide educational materials. Large predators have also made a comeback throughout the mountains of Europe and the grazing areas of eastern Europe. The Large Carnivore Initiative for Europe acts as a clearinghouse for many research and national groups throughout this wide area. Practical information and research can be found on the Web sites of all of these groups and governmental agencies (see Appendix, 213).

We were also told that guarding ability is instinctual, requiring no training or guidance by the human owner; that dogs will just stay with their animals (no need for fencing!); and that we could breed any two livestock guard dogs together, since they are all the same, whether they were born in France or Turkey. And we were told to expect a high failure rate in terms of dogs that won't stay with the flocks, dogs that chase stock, and dogs that are killed by cars or disappear.

That last, unfortunately, proved to be true — mostly because we didn't know what we were doing. Today, we are steadily gaining more knowledge not only of livestock guard dog breeds, but also of methods of use, problem solving, and selecting for proper traits.

One important thing to remember is that livestock guard dogs traditionally watched the flock in the company of shepherds. The shepherd and his dogs took the flock out to graze during the day and brought it home at night to a protected area. Many flocks were taken up into the mountains to graze during the summer months, in a process called *transhumance*. Nomadic groups traveled with dogs as they moved their flocks on their journeys.

In all of these cases, dogs watched over the flocks, but the shepherds were nearby both day and night. Puppies were born into the sights and smells of the barn or among the flock animals. Older dogs and shepherds were able to model desired behavior and correct inappropriate behavior in young dogs. The livestock were not afraid of the dogs that guarded them.

We are asking livestock guard dogs to do something much harder in North America. We often ask them to function as independent guardians day and night. In addition, we are often asking them to do this without any training and guidance from an older, experienced dog, and with livestock unaccustomed to these dogs.

In spite of these early problems, many studies found that a majority of owners reported that their dogs were excellent or good at preventing predation and confronting threats from coyotes, bears, wolves, and mountain lions, as well as smaller predators. Although purchasing and caring for a dog can be expensive compared to a donkey or llama, producers were able to show that livestock guard dogs definitely saved them money overall. Other advantages include reduced labor and feeding for stock that no longer required night penning, the ability to use pastures despite heavy predation pressure, decreased use of less desirable methods of predator control, and overall protection of farm property.

Which Breed Is Best?

Studies that attempt to show which breed is best at guarding have been far less conclusive. At first there were few breeds to choose from, the number of crossbred dogs in use lessened the predictability of behaviors, and the owners lacked traditional knowledge about using or training their dogs. The old charts or lists found in early articles and government publications are seriously outdated. Most studies were done with breeds that were available or imported during the 1970s and early 1980s: Great Pyrenees, Komondor, Akbash, Anatolian, Maremma, Sarplaninac, and Kuvasz.

Information from university and government researchers suggests that Great Pyrenees (often referred to as Pyrs) are less likely to be aggressive to people, stock, and other dogs than Akbash (no plural form), Anatolians or Komondorok (the plural of Komondor). Pyrs mature more quickly than Komondorok and Anatolians. Komondorok are more likely to bite people. Akbash are more effective at deterring predation and are rated high for aggression toward predators and displaying attentiveness, trustworthiness, and active behaviors.

Akbash Dogs effectively deter predation through their attentiveness, trustworthiness, and active aggression toward predators.

Be very cautious about using studies from twenty to thirty years ago as a guideline when selecting a dog. Behavior can range widely within each breed. And as compared to a couple of decades ago, we have many more breeds to choose from, some carefully selected from working stock in Europe or central Asia. Most importantly, success is more likely with a pup from working parents that is socialized to stock at an early age. With the wider range of well-bred dogs available and the increased experience of breeders and producers using livestock guard dogs, a new, broader study of breed characteristics would be immensely valuable.

Behavioral Characteristics

Wolves are highly social animals that form strong bonds in their groups, a trait that was essential in facilitating the domestication of the dog. But the ability to live in social groups is not limited to canines; it is one of the essential characteristics necessary for domestication of many kinds of animals, such as horses, cattle, sheep, goats, and llamas. The dog and the housecat are the only domesticated predators (and some experts question how domesticated the cat actually is). We make use of the canine

ON THE FARM

Dan and Paula Lane
Bountiful Farm, LeFlore County, Oklahoma

In the wooded hills above the Poteau River Valley in Oklahoma, Dan and Paula Lane raise Boer goats and Great Pyrenees livestock guardians. The twelve adult dogs deter a large coyote population and the possibility of attack from numerous free-roaming domestic dogs, in addition to cougars, bears, bobcats, wolves, eagles, and hawks.

The Lanes believe that the Great Pyrenees is the livestock guardian that is least aggressive toward people. They "love their personalities, their appearance, and most everything about them" but, most importantly, feel that "they do their job well, day in and day out." Bountiful Farm is also home to a breeding program that seeks to select and combine good working traits with the larger stature of the Great Pyrenees found in Europe.

Dan and Paula move their dogs around the various pastures so that they are accustomed to any situation in which they find themselves. Generally they work in pairs, but at times as many as five dogs will work together. The Lanes continue to marvel at their capacity for teamwork.

They cannot stress enough the importance of monitoring young and adolescent dogs, which still have the possibility of misbehaving. Bad behavior must be stopped immediately, the Lanes say, and the dog needs to be kept in an environment that does not allow for the possibility of that particular misbehavior. Other than this vigilance by the owners, most training is done by the older dogs. The Lanes do caution new owners to respect the working relationship with their dog and to avoid making their livestock guardian into a pet rather than a partner. They find that this remains a bit of a challenge even for them. With multiple breeding dogs, however, their biggest problem is keeping intact dogs separated on a working farm.

The Lanes have worked hard to gather much practical advice on raising livestock guardian dogs and post it on their Web site (www.bountifulfarm.com), and they share their knowledge in presentations and respond to questions from an e-mail list as well. Their advice to a potential owner? "Do extensive homework about the breeds, [about] possible breeders, and about using livestock guard dogs effectively — *before* you get your first dog."

aptitude for social living and communication when we raise and train livestock guard dogs.

In livestock guard dogs, the period of social development from two to sixteen weeks of age can be critical in forming a puppy's lifelong social bonds with the animals it will come to guard. The dog does not think he is a sheep, contrary to what is often said. The livestock guard dog knows he is a dog and displays canine social behaviors, rather than predatory ones, toward the sheep or other animals, just as he does toward humans, if he is properly socialized to them during the right time.

Domestication is a complex process, affecting both physical and behavioral traits. Because mammals change so much in shape as they grow, they have great potential for developing human-directed differences. In effect, humans can stop the physical and even behavioral development of mammals at different stages. This is very evident in the dog, expressed in the more than 600 widely varying

Livestock guardian dog pups like this young Great Pyrenees possess a longer period of social bonding than most other dog breeds.

breeds found around the world, some with very wolflike appearance and behavior, and others with extremely puppyish appearance and behavior.

The famous experiment in fox domestication conducted by Dmitry K. Belyaev and his research group at the Institute of Cytology and Genetics in Siberia, begun in the late 1950s, demonstrated that selecting for tameness alone caused important and profound changes in development, both physical and behavioral. Belyaev began his experiments in the hope of developing a more docile fox for use in commercial fur farms. Belyaev selected his foxes strictly for tameness, but the resulting litters soon displayed a startling range of new characteristics, including a delayed development of the fear response, changes in coat color (such as white patches), floppy ears, curled tails, shorter and wider snouts, earlier sexual maturity, longer breeding seasons, and larger litters. Now, after some 30 to 35 generations, the selected foxes are docile and eager to please, and they seek human attention. They no longer have desirable pelts, but the continuing experiment has taught scientists a great deal about the process of domestication.

One aspect of canine evolution that can be manipulated through selection is "neoteny", or the retention of juvenile traits in an adult, including behaviors such as submissiveness, attention seeking, begging for food, waiting for the adult to return, playing, and barking. Another of these traits is a delay of the fear response to strangers. In wild foxes this occurs at about six weeks, while the Belyaev foxes now show it at nine weeks, and domestic dogs show it between eight and twelve weeks or later, depending on the breed. A delayed fear response is important because it allows the puppy more time to form social bonds to humans or other animals that are not dogs. Neoteny is important in explaining the basic nature of the livestock guard dog breeds, most of which look like big overgrown puppies even in adulthood. Moreover, these big puppies with their floppy ears and curly tails don't look much like wolves, and the sheep have come to accept these "not-wolves" among them.

PREDATORY BEHAVIOR OF DOGS

We have also learned that dogs have several behavior systems, each regulated by biological processes such as hormonal secretions, motor coordination, and sensory perception. These behavior systems include fear, investigation, play, and submission. Each of these behavioral systems has its own rate of development, and these systems vary among the breeds or types of dog. One set of motor behaviors is predatory. Predatory behaviors occur in this order: *orient–eye–stalk–chase–grab–bite–kill–bite–dissect*.

The best adult livestock guard dogs don't display any of these predatory behaviors. Some livestock guard dog puppies display developmental predatory behaviors such as chasing or grabbing. These behaviors appear at five to eighteen months of age, but they can be extinguished if the critical bonding/socialization period has occurred successfully and the pups have formed social behaviors toward livestock. If a human or an adult working dog stops the undesirable behaviors when they happen, they will most often disappear by the time the puppy is an adult. A puppy practicing predatory behaviors with stock is not really displaying true aggression toward the stock. Rather, he is attempting to play with his companions, and inappropriate play can be stopped just as you would stop the inappropriate behaviors of a puppy in your house or with you.

Different breeds display different behavioral systems. For example, Border collies display their famous "eye" and "stalk" behaviors at a very early age. Sight hounds (such as the greyhound) specialize in the chase. Terriers are driven to catch and kill small prey. Protection dogs will grab on command. None of those breeds makes good livestock guard dogs, because those are all predatory behaviors.

INHERITED GUARDIAN TRAITS

Livestock guard dogs have been selected for centuries for low prey drive and the important traits of attentiveness, trustworthiness, and protection of stock. These breeds generally have a longer period of social bonding than the herding, hunting, or terrier breeds. This allows the livestock guard to bond with his companion animals. If you think it seems contradictory that good livestock guard dogs are aggressive with outsiders but highly protective of their stock, it may be helpful to think of this as the difference between hunting for prey and protecting pack mates.

Some adult livestock guard dogs that have never been socialized with livestock still make superb guardians because of the instinctive behaviors that come into play under the correct circumstances. However, most puppies will become good guardians if they are properly socialized and trained when young. Some will not — they may be too active or have too strong a prey drive. The fact is that in the past shepherds deliberately culled unsuitable dogs. We don't need to cull undesired puppies today. If they are identified early by breeders, puppies that display too much activity or too strong a prey drive can make excellent or even preferred companions or farm and home guardians. Some experts predict that 25 to 50 percent of puppies from any given litter will not prove to be good guardians; however, breeders who heavily select their dogs for good guardian traits have a much greater rate of success.

Understanding the complicated biology and development of dogs such as the livestock guardians gives us a whole new appreciation for these animal partners. It helps us select a good puppy for the job. In addition, understanding their nature helps us as we shape and train them to be good partners in caring for our farm or ranch animals.

> ### Deer Problems?
>
> **Livestock guard dogs** can keep deer out of your pastures and fields. In fact, studies at the National Wildlife Research Center have shown that livestock guard dogs are effective at keeping deer away from cattle. This may reduce the transmission of bovine tuberculosis from deer to cattle, a real benefit in those areas where this is an issue.

This pup is displaying desirable traits towards stock — calmness and submission.

4
SELECTING AND TRAINING YOUR DOG

Before choosing a livestock guard dog you need to ask yourself several questions. Do you have an immediate predator problem, or are you facing potential or increasing problems? Do you want a male or a female dog? Do you plan on using this dog as a full-time livestock guardian or as a farm guardian? Are you prepared to do the intensive training and commit the time that is required to produce a good dog?

Do you have a basic understanding of how livestock guardian dogs work and what their strengths and limitations are?

Once you have answered these questions, you can start thinking about the right breed of dog. Choosing a breed requires that you do some research and carefully consider your needs. Although no one breed is better than another, one breed may fit your particular needs better. As you read about the different breeds and visit them in person, pay careful attention to their overall breed traits, despite the differences in individual dogs within a breed. Consider requirements such as your farm or ranch's physical situation, your husbandry style, the types and numbers of predators in your area, your livestock, the neighbors and other people who visit your property regularly, your climate, your interest in grooming a dog, the size of dog you need, and the cost.

If you are looking for a farm guardian, some breeds are more tolerant and people-oriented. The more popular breeds are generally more readily available and may cost slightly less. You may need to wait longer for a rare breed and pay slightly more. Please take the time to visit and meet any breeds you are considering, either at a breeder's home or at the home of someone who owns this breed. (See chapter 6 for descriptions of individual breeds.)

If you have an immediate problem with predators, an adult or late-adolescent dog would be your best choice, although reliable full-grown guard dogs are difficult to obtain. A good working livestock guard dog is highly valued by his owner and not likely to be for sale. Occasionally good working dogs become available when their owners sell off their livestock. Regardless, an adult working dog or a started adolescent dog is valuable, so you should expect to pay a substantial price for such a dog.

If you consider purchasing an adult dog, you must meet him and assure yourself that you can handle him safely. An adult dog should have a written guarantee for both his health and his working ability. You must have a secure area to keep the dog in while he adjusts his new home and animals. Some dogs are so bonded to their animals or territory that these changes are very difficult. Other dogs, particularly if they have been exposed to changing livestock and territory, are more malleable and will make the adjustment over time. The most successful moves are made between similar situations, whether fenced pastures or open rangeland, and with the same type of stock.

How Many Dogs Do You Need?

For a single group of animals in one properly fenced pasture you will probably function very well with one dog. A general guideline is that one dog can guard 50 to 100 animals on less than 20 acres, but this is highly variable. If you divide your stock into various small pastures, you may need to provide access for your dog between these areas, or you may choose to place an individual dog with each group. Alternatively, you may choose to use a dog only in the pasture with the greatest predator pressures or

Rough or large pastures or the presence of large predators or packs of predators will necessitate the presence of more than one livestock guardian dog.

problems. Factors that may dictate your need for more dogs include animals that do not flock well and spread themselves out over your pasture, rough pastures with lots of cover for predators, and large predators or packs of predators in your area. Be aware that a single dog simply cannot handle a mountain lion, a bear, or a pack of coyotes or wolves.

The use of very large pastures or open rangeland generally calls for multiple dogs just to cover the territory and monitor the potential threats. Some of these situations leave the dogs in remote locations without human contact for several days. In these cases you need very independent, physically active dogs. Remember that traditionally in situations where stock grazed on open land, one or more shepherds remained with the dogs. That is still the ideal situation. Even in very large operations, the best management is to make a daily visit to the dogs and stock, checking on the welfare of both. The most successful ranchers in these conditions use older, experienced dogs to train the adolescent pups.

You may need to experiment with combinations of dogs to get the individual dynamics right. Pairs generally work very well, especially those of opposite sexes. One or both should be altered. If you need more than two dogs, neutering is a must. Older dogs usually accept very young dogs without problems, reducing the potential for conflicts as the younger dogs grow up.

Purchasing a Puppy

A livestock guard dog puppy is easier to find than a reliable adult or a started adolescent, because it takes time to raise and nurture a good guardian. Good behavioral instincts, careful breeding, proper exposure to stock, and correct reinforcement of appropriate behavior are the best guarantees for a working livestock guard dog. Once you have a properly trained adult dog it is easier to raise a second puppy, as he will learn a great deal from your first dog.

Either a male or a female will work well. Livestock guard dog research has shown that both sexes guard equally well, especially if they are neutered. Intact dogs in general are more likely to roam. Bitches are distracted from their duties if they are in heat or raising a litter of pups, while intact male dogs are usually slower to mature and more likely to be aggressive to other dogs as adults (though neutering does not reduce aggression toward predators, contrary to what you might hear). It is much easier to keep dogs of different sexes if you plan on eventually owning and working more than one livestock guard dog, as dogs of the same sex are more likely to fight even if they are both neutered or spayed. However, if

It is easier to buy a livestock guard dog as a pup, but he will require time to mature and will need training.

this is your first working dog, you should preferably begin with one puppy rather than two.

When you have determined the best choice of breed for your situation, you need to find a breeder and choose a puppy. The best place to locate a breeder is through a breed club, if one is available. Members of breed clubs generally subscribe to a code of ethics in both breeding and dealing with customers. They also screen for serious health concerns. They will encourage you to ask questions and provide support as you make your decision. Look specifically for folks who breed or place working dogs. Ask for references to owners of working dogs bred by this breeder.

If you know people with guardian dogs, ask them about their experiences with breeders. Dog magazines, livestock journals, and the Internet are other sources of breeders, but do your homework. In any case, do not be surprised if a breeder quizzes you or asks you to complete a questionnaire. Most breeders of livestock guard dogs are very concerned about finding good and appropriate homes for their pups. You might need to provide references as well. You may find yourself on a waiting list for pups and may need to place a deposit on your pup when he is born. In addition, most breeders will not sell you a puppy without a mandatory spay or neuter clause in the contract.

It is very important that you visit the parents of any potential pup. If this is not possible, you should arrange to visit siblings or other adult dogs of this breed. The breeder should be willing to put you in touch with satisfied customers who will introduce you to their dogs and answer questions. This is especially important if you have never owned one of the livestock guardian breeds.

Breeders should guarantee in writing the dog's working instincts and its basic health. Breeders may offer a free replacement if the dog does not work out or develops serious genetic health problems, although it is your responsibility to train and feed the dog properly. The parents should be free of genetic defects that impair the dog's ability to work, and they should be screened for hip dysplasia and other potential problems in that particular breed, if any. Puppies should have their appropriate shots and be wormed regularly. Ask what sort of support the breeder can offer as you raise your pup. Advice from someone who has working dogs is valuable, and many breeders are excellent mentors. Ask whether the puppies are being exposed to livestock. Good breeders socialize their pups to new people and situations, as well. The breeder should know the different personalities and behaviors of the pups well enough to help you make a good selection. The breeder may recommend a specific puppy to you. Conscientious breeders will often request that the dog be returned to them if you are unable to keep him for any reason.

Finally, make sure you receive a bill of sale or a sales contract, registration of the pup in your name even if you don't plan to breed him, and a clear understanding of how this dog will be picked up or delivered. Many livestock guard dog breeds are uncommon in North America, and you may need to have your pup shipped to you from a long distance. Buyers are generally responsible for any shipping costs and any necessary health certificates.

WHAT TO LOOK FOR IN A PUPPY

How do you select a good potential working pup? Real working parents or parents who have sired proven working dogs are the best source. The ideal situation would be a litter of pups already placed with stock so that you could observe their behavior. Look for the following characteristics:

- **Activity:** Dogs with lower activity levels are usually easier to train than highly active dogs. However, if your livestock guard dog will need to guard very large pastures, a higher activity level may be desirable.

Qualities to look for in a potential guardian include nonthreatening signals such as avoiding eye contact and dropping to the ground in the presence of stock.

- **Prey drive:** Some pups will already exhibit very low chase or prey drives, which you can test by throwing a small object.
- **Temperament:** Avoid pups that growl, bite, or struggle fiercely when you handle them. Look for a pup that is interested in you but not overly aggressive, clingy, or fearful. If you are looking for a full-time guardian, you need a pup that is a more independent-minded problem solver, and not dependent on human companionship. Look for a pup that walks off by himself, hangs back from the pack, or curls up and sleeps apart from the pile of puppies. Remember that you are not looking for the best pet or companion dog.
- **Pain threshold:** Avoid a pup with a low tolerance for pain, which you can test with a gentle pinch. Working dogs need to tolerate pokes and prods by livestock.

Obviously you should avoid a sick, lame, or unusually lethargic pup, but do not be put off by minor issues such as incorrect coloring or markings. Pups that might not be acceptable show or breeding prospects can make excellent working dogs. Avoid pups with pink pigment in eyelids, nose, and lips if you live in a sunny area. Repeated sunburns can lead to lesions or skin cancers. Be cautious of unusually large pups, which may suffer from orthopedic problems as adults.

If you are able to watch the pups interact with nonaggressive stock, look for a pup that is curious but somewhat cautious. He may or may not make eye contact with the stock; either is fine, though avoiding eye contact is preferable. Avoid pups that bark, jump, or bite stock even if they are accidentally stepped on.

Older pups should be submissive and calm when they are exposed to stock. Signs of submissive behavior include walking instead of running up to stock, dropping to the ground or rolling over, lowering the head and tail, and licking at the mouths of stock. Look for a pup that chooses to sleep next to stock, even through a fence.

Many breeders will help make this important selection for you. They observe the pups as they grow

Screening for Hip Dysplasia

At 16 weeks of age pups can be X-rayed and evaluated with the University of Pennsylvania Hip Improvement Program (PennHIP) technique to screen for potential hip problems. This testing can be important in breeds with very high levels of hip dysplasia or when you are investing in a valuable or potential breeding-quality dog.

and interact with each other. They may administer basic temperament tests as well. If your puppy is coming from a distance, you may need to trust your breeder to make a good choice for you. Current research indicates that pups need to stay with their mother and siblings until eight to ten weeks of age in order to learn how to play or interact with other dogs and develop proper bite inhibition. If your pup is receiving good livestock experience and you are a first-time owner, consider extending this time through an arrangement with your breeder. However, if the pup is not living with stock, you should bring him home and begin the bonding/socialization process.

Cost of a Puppy

How much should you expect to pay for a carefully bred, registered puppy from health-screened parents? A price tag of $1,000 to $1,600 is common for many breeds, puppies in great demand, or proven adult dogs. Breeding-quality or rare-breed puppies may cost $1,600 and up. In the more common breeds puppies may be available for $750 and up. Less expensive dogs may be available from unregistered parents or those without health screens. Occasionally high-quality working pups may be available at a moderate price from an owner who breeds primarily for his own need for working dogs but may have a few extra pups to sell to good homes.

Be extremely cautious of ads for puppies priced at under $500. Breeders simply cannot buy good breeding stock, perform the necessary health tests, give proper medical care, and provide good food for growing puppies for this low price.

Finally, this is a blunt statement but true: you cannot buy a puppy out of a box at an auction, throw it out in a field, and expect it to be good livestock guard dog. You might be lucky, but the odds are definitely against you.

THE ISSUE OF CROSSBRED DOGS

Tremendous numbers of puppies that are the result of two or three different livestock guard dog breeds being bred together are offered for sale. In addition, many puppies result from unions between a livestock guard dog and another "good" farm dog of another breed. Sometimes this is promoted as "healthy hybrid vigor." Technically, hybrids occur between members of two different species, such as horses and donkeys, not two different breeds of the same species. More importantly, crossing two breeds does not guarantee that the puppies will display the traits you want.

Crossing an alert and highly responsive livestock guard dog breed such as the Caucasian Ovcharka with a calm, placid breed such as the Spanish Mastiff will not result in puppies whose behavior falls between the two extremes. Some puppies may be like their mothers, others may be like their fathers, and others may have completely unpredictable behaviors. In addition, you can't use appearance or phenotype to determine behavior or temperament. A puppy may look like one parent and act like the other parent.

In fact, livestock genetics has shown that crossing breeds actually disrupts inherited behavioral sequences and eventually leads to their extinction or loss. More immediately, even in the first cross you lose the predictable behaviors that were present in the two parents. Consequently, crossbreeding must be evaluated very carefully.

Despite the common misconception, crossbreeding does not prevent genetic health issues either. If one parent has hip dysplasia, odds are that a high percentage of the pups will as well. Crossbreeding has no effect upon this. In fact, purchasing a crossbreed is an even greater gamble if you know nothing at all about the parents' and the grandparents' hip status.

The proof of all this lies in the shelters and rescue organizations that are filled with hundreds of crossbred livestock guard dog dogs that have failed at their job, run away, or been dumped. It cannot be repeated enough: you get what you pay for. An inexpensive puppy will cost the same to feed, medicate, and care for as a carefully bred dog from a breeder who performed health screening on the parents. Meanwhile, you will invest many hours training and socializing this puppy to your stock, perhaps only to find out that he is poorly qualified to be a livestock guard dog or has crippling hip disease.

Yes, there are some excellent crossbred livestock guard dog working on farms and ranches today, but frankly, your odds of failure increase when you choose a crossbred puppy. If this is your first livestock guard dog, make your job somewhat easier by choosing a puppy from a more predictable background. You simply cannot expect consistent behaviors from two very different parents. In addition, even well-intentioned folks who crossbreed their livestock guard dogs probably do not perform expensive health testing on the parents, since they are not able to recoup these costs by selling registered or purebred puppies.

Finally, do not be tempted by a cross between a herding dog and a livestock guard dog, even though these puppies are common, since both types of dogs often live together on the same farm. This is the worst possible combination in terms of a working livestock guard dog. The pups will probably have the drive to chase and herd stock combined with great size and power. In my opinion, this type of dog should not even live on a farm with stock.

Working Through Rescue Groups

Volunteers run most rescue groups, which may be devoted to a specific breed, crossbreeds of related groups of breeds, or any type of dog. If you are interested in obtaining a rescue dog for livestock guard dog work, you should begin with groups affiliated with or formed for the livestock guard dog breeds. The first place to look is the Web site of the national breed club. These groups are knowledgeable about livestock guard dog behavior and they know how to evaluate potential rescues for good homes.

Be cautious of groups or individuals who rescue many animals, regardless of breed. These groups, however well intentioned, probably lack the knowledge and ability to correctly evaluate a dog's potential as a guardian. This may sound harsh, but if you answer an ad for a free or inexpensive livestock guard dog, please be aware that the person you speak with may be motivated to color the truth, misguide, or even deceive you about the dog.

Reliable rescue groups usually have strict adoption procedures, which include an application, a reference from a veterinarian, a phone interview, and possibly a home visit. You may be asked to sign a waiver that releases the organization from any liability after you assume ownership of the dog. Specifically, you are responsible for this dog's future behavior with your other animals and any humans. During the home visit, rescue volunteers are looking to determine whether the dog will have a safe home with reliable fencing and an escape-proof pen with provisions for shelter, food, and water. The volunteers also need to find out which family member will make the commitment to closely supervise and train the rescue dog for as long as necessary.

There is usually a significant adoption fee for a rescue dog, which covers the expenses required to transport the dog from his original location to his foster home and finally to you. This fee also covers some medical care, feeding, training, and socialization.

ADOPTING A RESCUE DOG

A mature dog that was originally a family pet is another option. Many livestock guardian dogs originally purchased as pets are turned in to rescue organizations because they bark too much, shed too much, or are hard to control in an urban or suburban setting. Some of these dogs will actually be happier with a job to do and some space to stretch their legs. Given careful support and time to adjust, a dog with good genetic instincts may turn into a good farm guardian and perhaps a good livestock guard dog as well.

Look for a full-grown dog that is comfortable around people but not clingy or fearful. The dog should be accustomed to living outdoors but not prone to escaping, wandering, or chasing other animals such as cats. Most importantly, any potential dog should be tested for his reaction to stock; you are looking for calm or submissive behavior here. Initial curiosity should be expected. Dogs with low activity levels are much better than high-energy dogs. Again, make sure you can handle this dog and that you are comfortable having him around your family.

Some rescue groups will not place dogs in working homes at all, some groups will do so only with good preparation and evaluation of the dog and the potential new home, and some groups are much too eager to give you a dog. Please be aware that some sellers of adult dogs and some kindhearted people who work in rescue organizations may misrepresent problem dogs. In fact, experienced livestock guard dog owners who regularly accept rescue dogs to retrain say their greatest problem is misrepresentation of the dogs, which sometimes have serious or even dangerous behavioral problems. Some of these problems may not appear right away.

Behavioral problems are generally more serious than medical problems. Any dog that is being re-homed, for whatever reason, will be insecure or prone to separation anxiety at first. If malnutrition was a problem, the dog may be overly protective of his food. Neglect and lack of training or socialization may also produce a dog that is overprotective of his food or toys. He may be fearful of children or strangers. He may lack basic manners and housetraining. He may have bad habits that arose out of boredom, such as barking, chewing, digging, and other destructive behaviors. If the dog was removed from his litter at too early an age, he may not have learned how to play and interact with other dogs or developed proper bite inhibition. Finally, any livestock guard dog that is not neutered and not socialized often shows dominant aggressive behavior, directed toward other dogs, other animals, and even humans. A dominant and independent dog is best suited for a very experienced owner, rather than a livestock or family farm situation.

If a dog has already failed at livestock work, the chances are very small that he will improve in a new situation. It is possible that he will do better with different stock, a different pasture situation, or a different handler, but he will bring with him any potential problems with chasing or killing stock, escaping pastures, aggression toward humans, and so on. The proper raising of a livestock guard dog pup is critical. When you rescue a poorly raised pup, you will probably not be able to make the behavioral repairs necessary to make him a reliable livestock guard dog. And this is definitely not a job for a first-time livestock guard dog owner.

In addition, as opposed to a puppy from a conscientious breeder, you will probably have no knowledge of the rescue dog's health history or his parents' health history. Such knowledge is especially important in the case of medical conditions that may develop over time and prevent the dog from working, such as hip dysplasia. Many rescue dogs have internal and external parasites, although these conditions should be treated by the rescue organizations.

ON THE FARM

Stuart and Bob Richens

Banks Mountain Farm, Hendersonville, North Carolina

On a farm located up a winding road in western North Carolina, Stuart and Bob Richens raise Boer goats for meat, breeding, and show. Their primary predators are coyotes and feral dogs, but there are also black bears in the neighborhood. The Richenses utilize a rotational grazing system with both portable and electric fencing. They welcome many visitors to their farm and enjoy taking their animals, often accompanied by a livestock guard dog, to farm and goat shows.

Five adult Kangal Dogs presently work as full-time livestock guardians at Banks Mountain. They are moved around to suit the needs of both the dogs and their owners, especially as the groups of goats change throughout the year. Sometimes the dogs work alone, and at other times they are set to work in pairs.

The first livestock guard dog at Banks Mountain was a crossbreed. This experience, along with their observations of different dogs on other goat farms, led the Richenses to the decision that in their particular situation they needed a breed that was an effective guardian and also possessed a steady temperament. After some research, they settled on the Kangal Dog. The dogs have been so successful that the Richenses now breed the occasional litter of pups.

Pups are born and raised with the goats. When the puppies are still very young Stuart keeps them in a puppy pen made from livestock panels inside the pasture, but she has observed that some pups exhibit an early and persistent determination to stay with the stock.

Stuart believes that the best situation for any pup, as soon as he is ready, is for him to be kept with both an older dog and a couple of reliable adult animals. Choosing the stock wisely is very important, she says; you want a mentor and a tutor, but not a bullying, frightened, or very young animal. She often uses two bucks to act as mentors for the pups.

The puppy pen continues to function as management tool, especially if she is not around to watch the pup closely, but Stuart also believes it is important to have the pup outside the pen as much as possible and to watch him closely for signs of being comfortable and responsible with the stock.

Stuart actively works to create a bond with her working dogs, believing that this bond is just as important as the bond they forge with their stock. Her mantra is "Socialize, socialize, socialize!"

According to Stuart, the most important skill you can develop is to pay attention to what your dog is telling you. This way you can be proactive and avoid many problems. Observation will also tell you what your dog can handle. It is important not to expect more of dogs than they can deal with, especially when they are young dogs.

Her best advice to new owners is to obtain a pup from a good breeder who observes pups for good working traits. In addition, as she says, "Listen to your mentor!"

Stuart's biggest challenge in working with her dogs has been working out the "doggie dynamics." It is important to be aware of changes between the dogs, because, as Stuart says, "They do change!"

Raising and Training Your Puppy

In the past some livestock guard dog users advocated raising a puppy away from almost all human contact. Frankly, this is dangerous for both the owner and the dog. Dogs raised in this manner can become difficult or impossible to handle safely. All livestock guard dogs must be leash trained, accustomed to nail trimming and basic grooming, and receptive to handling by you and your veterinarian. Everyone needs a vacation and emergencies happen; your dog needs to be trustworthy around caretakers you may hire to mind your farm. This basic attention will not prevent the puppy from bonding or socializing to its stock.

When livestock guard dogs were first promoted in North America, there was a mistaken impression that these dogs worked alone. As we learned more about them and discovered how they worked in their native lands, we found that these dogs were most often in the company of shepherds. Either the dogs went out with the shepherds during the day and home at night, or the sheep, shepherds, and dogs camped out in the mountains throughout the summer. Dogs may have watched the flocks while the shepherds milked or made cheese, but in most cases the shepherds were not far from the dogs. At night when the dogs patrolled or slept near the penned or grazing sheep, the shepherds were close by and ready to respond if the dogs sounded an alarm. When we look at old pictures of shepherds and their dogs, we clearly see affection in their working relationship. We often ask our livestock guard dogs to do something more difficult here in North America: to live with the sheep and see the shepherd only once or twice a day or, in some cases, not for days at a time.

As we have gathered more experience and knowledge from working with our livestock guard dogs, we have learned that forging a strong relationship between dog and master will not interfere with the dog's ability to guard his stock. In fact, he will work better knowing that he has a leader in his job of guarding his stock. He will respond better to your praise and want to please you by behaving correctly. He will also respond to your displeasure. Moreover, your dog will be a better livestock guardian because of your continued training. Don't be afraid to be a shepherd and establish a strong relationship with your dog. Yes, your puppy needs to be socialized to his stock, but he also need to be socialized to you.

A good sign of appropriate submissive behavior in a livestock guard pup is rolling over or licking at the mouth of stock.

BRINGING YOUR PUPPY HOME

Do not buy a puppy to train as a full-time guardian if you do not already own stock or cannot borrow a couple of dog-wise but not overly aggressive adult animals for an extended time. Before you bring a puppy home, prepare an escape-proof pen in your barn or pasture where your puppy will be with or next to stock. Livestock panels, available at farm supply stores, are excellent for this purpose. The pen can be constructed with T-posts so that you can move it or expand it as needed. Livestock panels also work for tops and floors if your pup is a climber or digger. Your puppy or dog will also need shelter from the weather if he is outside. Eventually you may need a stock-proof feeding station for your dog, and the pen you build may serve this function as well. An eight-week-old puppy needs a space of only about 16 square feet, although he will need more room as he grows or if he is housed with stock.

It may be necessary for this pen to be out of sight of your own house and yard, where the pup would regularly see you, family members, or pet dogs. If your pup came from a situation where he was raised with stock, he will feel more comfortable and content with them. If he came from a different situation, he may cry for you if he can see you or your pets, and this may interfere with his socialization to stock. You will need to assess your own situation and the pup's experience.

Do not bring the puppy into the house when he arrives. Do not let your pet dogs play with him. If you do, you will make his eventual separation from them that much harder. If he has traveled a long distance, he may need small drinks of water and a bath. If at all possible he must be placed in his new home and receive his attention and feeding there. Experienced users often say, "All good things come in the pen or with stock." Visit him and give him attention in his pen or in the area with his stock. Don't be afraid to spend some time with him, but don't respond to his calls or condition him to demand your attention. Introduction to a leash can also take place in the pen. He will need something to safely chew and play with in his pen as well.

PROPER SOCIALIZATION

Your dog needs to be socialized if he is to live a useful life. He needs assistance in evaluating potential threats and developing his own measured response to threats. In most cases you will have purchased a livestock guard dog to protect your stock from predators, not humans. You want your dog to learn to accept new people and dogs when you introduce him to them. He will still remain vigilant against real threats or people who approach your home when you are absent.

Do not worry if your young livestock guard dog does not exhibit protective behavior, and do not purposefully encourage aggression in your pup or dog. Many livestock guard dogs do not become protective until they reach maturity at 18 to 30 months of age. Your friendly pup may change quite drastically as he becomes an adult.

While he is still a pup, your dog should become acquainted with all of the people who work or regularly visit your home or farm. He should meet your pets and other working dogs. How you handle these canine introductions is a decision you will need to

Livestock panels are an excellent material for creating puppy pens.

base on your particular situation. Some livestock guard dog users firmly believe that all other dogs should be kept out of the livestock guard dog's working space. Whenever you use a herding dog, you do need to leash your young livestock guard dog, tie him, or put him in his pen. You do not want him to learn to chase or herd stock, but you may want him to calmly accept you working with your herding dog.

It may be difficult to take your livestock guard dog off your property to be socialized with people, although occasional visits to the vet and other places are beneficial. You should introduce your dog to everyone who visits you. Your pup may be extremely friendly and seek attention from strangers you bring to his pasture. If not, leash your dog and allow him to observe visitors while you converse. Many livestock guard dogs are quite aloof to strangers, although your dog will come to see your calm presence and the leash as a signal to accept any new person. If your dog continues to bark or growl at a visitor, you need to work on teaching him to quit this undesirable behavior. Set up and encourage situations in which your dog can be calm and friendly, and then reinforce that behavior as much as possible. It is much better for your dog to learn good manners as a pup than as an older, stronger dog.

If you have access to a dog crate, it can be very valuable to train your pup to accept short periods of time in a crate, even if he will be a full-time working guardian. Situations may occur later in his life when he might need to be crated, such as natural disasters, serious injuries, or overnight treatment at a veterinarian's office.

Introducing Livestock

There are different ways of introducing a puppy to his new stock. Good breeders should be able to give you advice that they have found helpful with their own dogs, and you should follow this advice. You may need to choose and adapt your process to the temperament of both your pup and your stock. Some dogs take a long time to become reliable with stock; others become reliable in a short time. Although we use the word *training*, we are actually *shaping* and *encouraging* the livestock guard dog's natural instincts. When we use the word *bonding*, in many ways we actually mean *socializing*.

One method is to place lambs or kids in the pen with the puppy. Bummer lambs or kids that need to be bottle fed regularly might be a good choice. You will need to monitor their interactions very carefully so that the growing pup does not play inappropriately with the young stock. Animals that are smaller than 20 pounds are probably too small to be companions for a growing puppy. This method generally works better for pups raised with stock from birth. Please be aware that many experienced breeders oppose this practice, since they believe it increases the odds of the pup injuring or killing a

---

Training Puppy Pairs

If you are raising two livestock guard dog puppies together, they will benefit from being able to play with each other, and eventually they will work well together. However, you must give them separate attention and handling time so that they don't focus completely on each other. If you plan to have these pups working in separate pastures as adults, they must begin to spend time apart as they grow up. Be extremely cautious that two pups don't begin chasing stock together — pups tend to be braver when they have a playmate to back them up. Many breeders would recommend that you raise a single livestock guard dog before tackling multiples.

lamb or kid, even if by accident. Again, it is better to be safe than sorry.

Most owners place the pup in a pen with a couple of calm, mature animals. In this case the pup will need a place to retreat from them if they become too aggressive toward him. Overly aggressive stock can not only hurt the puppy but may also make him fearful of stock. On the other hand, skittish animals will reward any attempts your pup makes at chasing. If you have reliably calm, steady animals to pen with your puppy, the stock will discipline inappropriate behavior in the pup for you, and your pup will form a close relationship with his stock.

Another good choice is to house the puppy next to the stock, separated from them by a fence. When you feed him, as you'll need to do two or three times a day, spend some time visiting with him and begin leash and manners training. As he becomes comfortable with leashing, take him into the stock areas as you do your chores. Some people leash the dog to them as they work. Do not walk him outside of his eventual pasture areas. Do not allow chasing or biting — you must catch him in the act and interrupt the behavior immediately to make a correction. Verbal corrections will work with a sensitive dog, but many pups will need mild physical corrections (like those given by an older dog or experi-

If you do not have calm stock for your pup to socialize with, another good choice is to house the pup next to the stock separated by a fence or livestock panel.

Butchering

Do not butcher or kill livestock within sight of your livestock guard dogs, especially those animals he has protected. Some dogs find this very upsetting and perceive it as a threat toward their animals. Some dogs become extremely distrustful of the person who did the slaughtering. On the other hand, you can feed your dog bones and scraps from animals you raised. Your dog will not begin to kill and eat your animals once he has a taste of the meat.

enced stock animal) or time-outs back in their pen. As your puppy becomes more reliable, try leaving him for short periods with the stock while you can monitor him from a distance. This is also a good time to place him in a larger enclosure with a few experienced animals.

Depending on your pup's previous experience, his first birthing season, with its different sounds and smells, may be confusing for him. Although some livestock guard dogs behave appropriately even when they are immature, others may interfere with the birthing process. Your dog may try to mother the baby by licking it or preventing the mother from claiming it. The mother may become aggressive toward the dog, and the dog may in turn attack the mother. Your dog may also try to mother an abandoned or stillborn baby, so do not be too quick to assume that he has stolen the baby.

The best solution is to allow the leashed dog to watch several deliveries with you, learning not to interfere with the mother and baby. Rebuke him for any aggression or interference. Allow the dog to calmly observe mothers and babies together in lambing pens. Do not leave him unsupervised in a lambing pasture until you are confident of his behavior, which probably will not be until his second birthing season. He may consume afterbirth; this is natural for all canines, and he will not understand your displeasure with his behavior if this bothers you. Eating afterbirth will not lead to stock killing.

Pen your pup next to the poultry you want him to guard. As he matures, carefully supervise his experiences with them to prevent him from chasing or injuring the birds.

Guarding Equines

Livestock guard dogs are used to guard miniature horses with great success. Many owners also report that their full-size horses learn to be comfortable with livestock guard dogs. Exercise caution when introducing your livestock guard dog to horses, since the horses may panic and injure themselves in a natural flight response from a potential predator. Some equines are quite aggressive toward strange dogs.

Horses can also excite chasing in dogs very easily. Allow the horses and dog plenty of time to get to know each other through a fence. Be alert for the occasional horse that shows strong antipathy toward dogs, especially mares with foals and stallions.

DOGS GUARDING POULTRY

Livestock guard dogs were not traditionally used to guard poultry in their homelands, yet some owners have successfully socialized their dogs to birds. Others admit that despite their best efforts, some dogs just never become trustworthy around poultry.

Ideally, this process should begin very early. A pup should be penned next to the poultry you want him to guard. Periodically allow the pup to smell a bird you are holding, making sure it cannot peck the pup. Praise a good reaction, and scold any attempt to bite or mouth the bird. As the pup becomes more reliable, he should accompany you as you work with your birds. Do not allow any play or predatory actions from your dog, especially chasing. Verbal and physical corrections will be required. You cannot expect a young dog not to play, so you will need to channel his playfulness elsewhere. Continue to praise good behavior.

This process may go very slowly, especially if your birds are not used to dogs. Experienced users admit that you may lose a bird or two along the way, because birds are fragile compared to a half-grown livestock guard dog. Many livestock guard dogs are maternal toward tiny creatures, but unfortunately, even gentle licking can kill a small bird. Some livestock guard dogs do better with adult birds rather than hatchlings and young birds. Be cautious with adult ganders and toms, which may behave very aggressively toward dogs. You will also need to consider how you might construct nesting areas to discourage your dog from eating eggs. Besides the impact on your egg raising, eating raw eggs in large quantities is not healthy for dogs.

Your livestock guard dog will easily work as a poultry guardian in situations where the birds are securely fenced away from him. Farmers who raise large numbers of pastured poultry, including ostrich, emu, and rhea breeders, have created buffer zones around their pastures, enclosing them in two perimeter fences and giving dogs the run of the long narrow area between the two fences. Other poultry raisers fence in the areas around their chicken houses or coops, having the dog patrol for raccoons and other small predators outside the fenced areas. You can easily place a chicken tractor (a small, portable unit for raising young birds) inside an existing pasture guarded by a livestock guard dog, so long as the chicken tractor and any runs are fenced with a sturdy material that will resist an inquisitive paw. You may need to use a scare (electric) wire to discourage your dog from testing the poultry fencing.

Dealing with an Adolescent Pup

At about four months of age, the prime time for bonding or socializing with stock is passing just as the pup is becoming too big for a small pen. Your pup is entering his long adolescence, which might not end until he is two years old or older. Some breeds mature earlier and others later, but in any case this is a time for patience and vigilance. Even though your growing pup will soon be as tall as an adult dog, he is still a juvenile, and during this time several important habits must be formed. Notably, this is also the age at which dogs are often turned in to a shelter or rescue organization due to bad habits, or may even meet with a fatal accident — usually due to owner failures. We always need to remember that for many centuries young livestock guard dogs learned to work with their animals in the company of shepherds and older, experienced dogs.

When he is four to five months of age, put your dog out with stock in an area you can monitor frequently. Avoid young stock if at all possible. Ideally, this area should be in sight of your house. You may still need a pen to keep the puppy safe when you are unable to oversee his behavior. He may not be ready to be turned out with stock unless you are present. On the other hand, he may do well for some time and then regress. Every dog is different, and you will need to adapt to your dog's particular behavior.

If possible, try to vary your young dog's experiences. Feed him in different places, and move him around into different pastures and with different groups of stock and other working dogs. The more you vary his adolescent experiences, the more flexible he will be at adapting to changing situations on your farm.

It is appropriate for pups to be curious about new stock, but they should approach other animals cautiously, walking rather than running to them.

Adolescent dogs often get into trouble because their owners expect far too much of them. Adolescent dogs can be just like human teenagers, with sudden fits of irrational behavior and tremendous amounts of restless energy or boredom. Instead of giving up on him, you should be vigilant in observing your dog. Catching a small problem and stepping back in your training is far preferable to finding yourself with a large problem. Even large problems such as chasing livestock or biting wool can often be solved by going back to earlier training and experiences.

Basic Training Is Essential

Your livestock guard dog pup needs all the basic training that your housedog receives, although he will probably not attend a puppy class unless his future is as a farm guardian or family companion. A working dog still needs to learn to walk on a leash and the basic commands of *no, sit, stay* or *wait, drop it* or *leave it*, and *come* or some other recall signal. Your livestock guard dog must also learn basic manners: no mouthing, biting, or jumping up. He will be a very large dog, and these puppy antics will not be appreciated later. Some livestock guard dog owners like to teach their dog to jump up on command, but make sure this is only by your invitation.

Teach him to accept nail trimming, grooming, and medications, such as those for heartworm and parasite control. If you will need to take him to a veterinarian's office periodically, teach him to jump into a car or crate; a dog that won't cooperate with loading will become very difficult to handle when he weighs 100 pounds or more. It is also advisable to tie your pup occasionally for short periods, as there may be situations where you need to restrain him while strangers or herding dogs are working with your livestock, and you don't want him to panic.

He must also learn to come when called, though even with thorough training, do not expect your livestock guard dog to have a totally reliable recall, especially if he is in pursuit of a perceived threat. The best suggestion for teaching recall is the reward of something irresistible, like your attention or a delectable treat. Never allow your livestock guard dog off the leash or outside a fenced area unless you are in a remote location and have confidence in his recall.

BE THE LEADER

The important message you are giving your dog through training is that you are the leader or the boss. Your dog will respect you if you are the "granter of resources," which means *you* give him the things he wants, including food, attention, toys, treats, and even walks. You must also control his behavior. Ask him to sit before receiving his food or other resources, including petting. While he is eating, trade his food dish for another treat. Make him wait while you go through the door or gate first.

Your puppy should be tied occasionally for short periods so that he comes to accept this restraint as an adult when you need to work with herding dogs or have outsiders work with your stock.

If he invades your space, push him back with your body, not your hands, because your dog may see your hands as paws, which he uses to initiate play. In many ways, handling and leading a big dog is like handling large animals such as a horses or cattle. If you are familiar with leading livestock, you should use the same approach, insisting that he respect your personal space.

Livestock guard dogs are notorious self-thinkers and they have a low threshold for boredom, so you shouldn't expect an obedience star. Short lessons are better than long ones, and your puppy's lessons should be conducted in his work area. Make your training fun, interesting, and positive, as most livestock guard dogs do not respond well to harsh discipline. The most effective training involves rewarding your dog for good behavior whenever possible, rather than relying on punishment for being "bad." He may associate a punishment not with the behavior but, instead, the person or thing that he is upset about, which only exacerbates the problem.

Be careful that you don't inadvertently instill the very behaviors that you are trying to get rid of! A good example is yelling at your dog to stop barking. If he is rewarded by your attention, even in a negative way, he will probably keep barking to solicit it. Instead, ignore the barking and praise him when he is being quiet.

Be consistent with your command language. However, you don't need to demand that, for example, your puppy sit every time you feed him. In fact, you shouldn't, since you will just condition him to sit when he sees his bowl, rather than listening for your command. You want to reinforce commands, not condition him to certain behaviors. You don't need to be extremely physical or harsh, but you can be dramatic with both your praise and your scolding. As you read the individual breed descriptions, you will see that most experts with these dogs do *not* recommend using force or excessive punishment.

Most livestock guard dogs will shut down or even become more aggressive if treated this way. Don't attempt the alpha rolls you may have heard about. Don't confront any aggressive behaviors with your own assertive stares. Instead, you should look away from your dog and not reward your dog's attention-seeking. Be extremely cautious about roughhousing with your dog. You are raising a large, powerful working dog, not a house pet. As usual, it is better to be safe than sorry.

In most situations an owner's inability to control his livestock guard dog is due to a failure of leadership. As we've said earlier, you must demand that your dog respect you. Moreover, it is extremely important that your dog see *every* member of the family as a leader. If you are uncertain about your dog's behavior and/or how to train a dog with positive methods and strong leadership, consult the resource section for recommended reading.

PROPER USE OF COLLARS

Give each of your livestock guardian dogs a plain leather or nylon buckled collar with identifying information. Since tags are easily lost, many owners set into the collar a flat brass plate engraved with their contact information. These brass plates can be obtained from horse supply catalogs or tack shops. When two or more dogs are working together, it may be safer to use breakaway collars to prevent injury or even death if one dog's jaw becomes entangled in another dog's collar. Some owners use wide, flat, leather collars, which seem to prevent another dog's jaw from working its way underneath and becoming trapped. Studs can be embedded in a leather collar to prevent other dogs from chewing it. Livestock guard dogs should also have tattoos or microchips as backup forms of identification should their collars become lost.

Metal or nylon choke collars or slip collars are not usually appropriate for livestock guard dog breeds and should never be left on a dog that is working. These collars, which are used primarily to restrain or punish dogs, have been proven to cause tracheal injuries. If your dog has a very large neck, large dewlaps, and thick or long hair, the slip chain or choke collar is not able to perform as it was intended anyway.

However, we do need to face the reality that a large, powerful livestock guard dog is always a bit of a challenge. He may outweigh you or outpower you, especially if you have arthritis or an injury that limits your strength. A livestock guard dog is a self-thinker, and while you have him on a leash he may be confronted by what he thinks is a threat, or you might encounter a poorly trained or supervised strange dog. He can also be quite insensitive to his own physical pain. Whatever the situation, you must be able to control your dog in public.

Of course, a well-trained, highly socialized dog that enjoys a good working relationship with his owner may be walked and controlled with a simple buckle collar. A slip collar may work well for another dog. Everyone needs to make their training decisions based on their own dog, but you should be aware that many livestock guard dog owners choose a training collar or a head halter to help them train their dog.

Prong or Pinch Collars

The training collar, commonly known as the prong or pinch collar, is often misunderstood by the public and even by some dog trainers. It looks scary, but it is actually far less likely to damage your dog's trachea than a slip collar because it produces even pressure around the dog's neck. Less physical strength is needed to use a prong collar, and it works through self-correction rather than punishment.

Buy a collar with the largest-size links (you may still need to buy two or three additional links to fit a very large neck). You may be able to use a medium-size collar, but large and extra-large collars are available. The snap-on version is easier to put on than the standard training collar. The collar should be placed just below the dog's ears. The chain portion should be flat. At first attach the leash to both rings, which will prevent tightening. Some dogs will need the increased control afforded by using a single ring.

You should always attach the leash to the dog's leather collar as well; you may need a fail-safe tab or strap to connect them. Some dogs are able to progress back to a plain collar, while others will always need the prong collar in public or in stimulating situations. The prong collar is not appropriate for young puppies, which should learn their basic manners with only a plain buckle collar.

From left to right: a choke or slip collar; a prong, pinch, or training collar; a Halti or Gentle Leader. These types of collars should never be left on an unattended dog.

Head Halters

Head halters are available from two manufacturers: Halti and Gentle Leader. The Gentle Leader is made of thicker nylon and tends to fit a livestock guard dog better. The metal buckle model is more durable. Care must be taken in fitting the halter; you will need either a large or an extra-large size. Gentle Leaders can be used on young puppies. Be aware that the head halter can cause some dogs to shut down, become depressed, or strongly fight the halter.

CAUTION: With either the head halter or the prong collar, never let the dog hit the end of the leash with force. Be cautious about using either the head halter or the prong collar with an aggressive dog or fearful dog. Never leave a choke chain or a prong collar on an unsupervised dog.

Dealing with Predators

Do not expect your dog to be aggressive toward predators at an early age. Protective guarding behavior normally comes with maturity, although some young dogs and even puppies may exhibit protective behaviors. Do not be surprised if your dog acts uncertain or fearful at his first encounter with a predator. Be reassured that virtually all mature dogs will be aggressive toward threats.

It is highly likely that you will never see your dog engage a predator. Most often he will bark or posture to keep it away or deal with it when you are not present. However, you may hear the highly frenzied, ferocious barking that often signals this type of event. Some dogs actually call for backup assistance from you.

Be very cautious about interfering if you find your dog confronting a predator. You can actually be a distraction for some dogs, which might prevent them from doing a good job. If the predator is small, you may be stunned at the fierce and often rapid conclusion. There may be no time for you to intervene, since livestock guard dogs often snap the neck and let the animal drop to the ground. The best advice is to leave your dog alone. Your dog may be highly agitated. Do not attempt to restrain him. The most common response is for him to race around his area looking for more threats.

If your dog is confronting a large predator or group of predators he may need your assistance. When you arrive, he may welcome your help and defer to you, or he may not. If he keeps himself between you and the predator or is engaged in a fight, be careful not to put yourself at risk from either the predator or your agitated dog. This is his job, and the best course will be to let him perform it. Firing a gun or using an air horn may break up the fight if you are worried for his safety.

Causes of Failure or Loss of Dogs

A five-year study at the USSES revealed the following causes for livestock guard dog failure. The greatest causes for failure are actually people problems, such as not fencing in your dog securely or allowing him to roam.

Cause of Failure	Percentage of Total Failures
Hit by car	23
Malicious shooting	23
Health problems	18
Accident in field	9
Untrustworthy	4
Cause unknown	23

You can help prevent the loss of your dog by doing the following:

- Keeping your dog on your property
- Using signs and other measures to alert neighbors about your livestock guard dog
- Buying a dog from health-screened parents
- Preventing your dog from encountering poisons, snare, and traps

Proper Fencing Is Vital

The leading cause of death in adolescent working livestock guard dogs is being hit by a car. Wandering off and being shot outside of their property are two other primary causes of death in young livestock guard dogs. Escaping fences and wandering are also the leading reasons behind people's decisions to re-home young livestock guard dogs.

Your dog *must* learn that is impossible to get out of his fence. Once he learns that it is possible to escape, he will try repeatedly. You must remember that he hears, sees, and smells potential threats beyond your fences, and his instincts tell him to investigate or pursue them. Formal boundary training is not particularly successful with an independent-minded livestock guard dog. Most importantly, a livestock guard dog is a large and powerful dog that must be contained securely in today's litigious society.

Good predator-proof fencing, as described in chapter 2, will help keep a livestock guard dog in his place. Livestock guard dogs can easily jump over 3- or 4-foot fences, so fences must be built high. They can also climb over or dig under fences; electrified or scare wires placed low and high will prevent digging and climbing. Gates must fit tightly enough to prevent an escape. Solid gates or gates reinforced with woven wire are better than tube or panel gates. Electrified wire can be strung in front of the gate, on the pasture side, to discourage dogs from investigating. Dips, ravines, uneven ground, gaps, and waterways are all potential escape routes; see chapter 2 for advice on securing these areas. Remember to keep electric wires tight, using inline wire tighteners.

It can be handy for your dog to have access between various areas of pasturage, so you might consider creating a porthole or dog door for him. (See page 87.) Most sheep and goats cannot climb through a hole that is 14 inches to 2 feet off the ground. You can use an actual dog door, a piece of vinyl or rubber, or a piece of sturdy wire mesh attached by sturdy rings to a hole cut in your fence. Be sure to remove all sharp edges. Other owners have created agility-style ramps between two areas, although goats can often utilize these as well.

INVISIBLE FENCING

Invisible or radio fence is another option for reinforcing a poor fence or as a backup for any fence, although it is not recommended for puppies before they understand basic commands. In this system, the dog wears a collar that makes a warning beep if the dog ventures too close to the fence and then, if the dog does not stop, administers a small shock. You should introduce this system carefully, so that the dog does not associate the shock with anything

Your fencing must be designed to keep your guardian as well as your livestock contained. Invisible or radio fencing is an option for reinforcing a poor fence or as a backup for any fence.

Using an Invisible Fence

In designing your system, remember that the single containment wire must form a complete loop from the transmitter. If you must cross buried electrical, telephone or cable TV wires, do so at a perpendicular angle. Since twisted wire cancels the radio signal from the transmitter, you can intentionally twist wire to create areas where your dog can safely cross the invisible boundary. The size of the protection zone around the wire is usually adjustable at the transmitter. With some systems, you can keep your dog several feet away from the physical fence.

The fence can be installed across waterways by running the wire underwater through a length of garden hose or PVC pipe. Special waterproof underground splices are necessary if you need to join two pieces of wire together or to make a repair to a damaged wire; these are available from the manufacturer or most home improvement stores. Detailed directions for designing your fences are available from the manufacturers.

You can buy this wire more cheaply in bulk on large spools from wire suppliers than from the manufacturers of these systems. Check to make sure the wire you purchase is compatible with the system you buy. Generally, it is solid single-strand wire. You can make the system more rugged by using a higher-gauge wire, such as 16-gauge, rather than the 14-gauge wire that is standard with most systems.

Invisible fence wire does not have to be buried if you purchase UV-resistant outdoor wire that can be fastened to the fence. You can staple the wire to the fence, burying it one to three inches deep only where it crosses gates or other openings. You can also use sod staples to force it next to the ground. Nevertheless, it is best to bury the wire, since sheep and goats can chew on the wire and break the circuit. This is more likely to occur in smaller enclosures where animals graze through the fence, or with non-electric fencing.

You should also purchase a lightening arrestor or protector for the wall transmitter. The lightening arrestor will save you money in the event of a lightning strike near your fence, especially if you have the wire aboveground. These units are available from the manufacturers.

If you choose to use an invisible fencing system, you must pay regular attention to the system to make sure it is in good working order. Check the batteries in the collar units monthly, and replace them when they become weak.

Most livestock guard dogs will need the longer probes that come with the collar because they have extra skin or fur around the neck. You may need to shave a small area on your dog's neck so that the probes make contact with the skin. It is important to use the rubber insulators between the collar unit and the probes to provide insulation in damp conditions, and you must be careful that the probes cannot make contact with metal tags. Check the dog's neck often, since the probes on the collar unit can seriously irritate the skin if the collar is too tight. Once they are used to this system, many dogs can wear the collar loose enough to prevent irritation, since they are conditioned to avoid the fence when they hear the warning beep.

Check the collar unit periodically for cracking and loose probes. Spray-paint the black collar unit a bright or fluorescent color so that you can find it in a pasture if a dog loses his collar. If you do need a new collar, the manufacturers may sell you a reconditioned collar replacement unit for less than the cost of a new collar.

other than approaching the fence. Follow the training directions recommended by your system. Dogs that are already experienced with electric fence generally adapt well to this system. Dogs that are raised with this system rarely test their fences.

Invisible fence is *never* suitable as the only barrier for a livestock guard dog, since many dogs will run through the shock in pursuit of a perceived threat. For a working livestock guard dog the only recommended use of this system is to reinforce an existing fence. Invisible fence is good at preventing digging and climbing, since these activities take time to pursue. It is also very useful in areas of deep snowfall and drifting over fences. Invisible fence can be used to enclose areas as large as 25 acres, and with multiple dogs.

Proper Signage

It is very important that you clearly label the pastures where you keep a working livestock guard dog. You can purchase or make a sign like this. Talk to your neighbors and let them know you have a working livestock guard dog in your pastures. Many livestock guard dogs have been shot by well-intentioned neighbors or strangers who thought they saw a dangerous dog in with someone else's sheep.

Photo courtesy of Anatolian Shepherd Dogs International. Proceeds of sign sales support research efforts. See page 215 for contact information.

Introducing an Adult Dog to Stock

Although some experts believe it is nearly impossible to socialize an adult dog to stock, experience has shown many breeders and owners that this is not true. If an adult dog has good basic socialization to people, no serious behavioral problems, and good instincts, he can adapt and enjoy his new job. Be aware, however, that you may run across some adult dogs that are actually afraid of stock animals or have a stronger desire to be with people than with stock.

It is easier for a mature dog than for an adolescent to make the transition from a companion life to a working life, since the adolescent phase of a dog's life is often full of turmoil, while the adult dog has settled down. If you begin with an adolescent, be aware that you need to proceed very slowly and with great patience with all of the adolescent behaviors that livestock guard dogs can exhibit. Although the techniques described below are most often used with rescue dogs, they are adaptable to your own mature dog if you acquire stock later in his life.

If at all possible, before you bring an adult dog home, take him to a veterinarian for a health exam, and have him groomed and checked for problems such as abscesses, hot spots, skin infections, and mats. Groomers and veterinarians are used to dealing with nervous dogs. These visits will give you important feedback on how your new dog reacts to strange people and situations. This is also a good time for spaying or neutering, if needed.

If your dog is a rescue, you must have a very secure place to keep him for at least the first few nights, since he may be depressed or frightened. Do not place him with your stock or any other working dogs you own. His pen should be next to or in the pasture with his new animals. If the pasture is large, you should confine the animals close to the dog. Move the stock's feed or salt close to the dog's pen to encourage their interaction. Pet the dog when you feed him or walk him, and work continually on training him to accept your leadership, but do not take him into your house or allow your children or pet dogs to play with him. Experienced livestock guard dog owners suggest a minimum of two weeks of quarantine time while you and the dog get used to each other.

During this quarantine time, try to establish routines for handling and feeding the dog. Routines reassure dogs. Walk the dog on a leash around the perimeter of the pasture several times a day. Observe the interactions between him and your stock. If the dog is unfamiliar with this species of livestock or if the livestock is unfamiliar with guard dogs, it may take several days for them to become comfortable with each other. If the initial interactions go well, begin walking the dog on a line about 30 feet long, so he can explore but you can still control him. Correct any undesirable behavior with a simple *no*. Some owners allow the new dog to sniff each animal as it passes through a gate or narrow run. If you have lambs or kids, hold one and let him sniff it. Talk to him constantly, reassuring him that these animals are yours. This may sound silly, but experienced people swear that their dogs learn from these actions. Once you feel comfortable that things are proceeding smoothly, the dog may be released with the stock.

You must supervise the dog and the stock closely for several days. If the dog was a pet in his former home and is not used to stock, this process may take two or three months. If the dog shows any sign of chasing or harassing stock he must be stopped immediately and returned to his confinement pen. You may need to confine the dog in a small pasture with two or three older animals that will correct any improper behavior. In all cases, be very alert for the dog's behavior during new situations, such as birthing times, herding, winter feeding in closer confinement, working with other working dogs, and changing pastures.

Take your time introducing an adult dog to stock. Allow a minimum of two weeks and continue to supervise the dog and stock closely for several days after you begin leaving them together and at stressful times such as during kidding or lambing.

Introducing New Stock

When you introduce new animals to an established herd or flock, you should observe your dog's reactions and behavior toward them for a few days. Since he will be curious anyway, it is a good idea to introduce him to the new animals while he is leashed or to place the new animals in a nearby area. The easiest solution for you is to pen them for a few days in an area next to your dog and the old stock. If the animals are not used to dogs, the acclimatization will take longer. When the new stock are introduced to your flock or herd, some dogs may be upset at the inevitable squabbles between animals, while others ignore these disturbances. If your dog is upset by them, you may need to remove him for a short time until things settle down.

Introducing Additional Dogs

If you have an adult guard dog already, do not immediately put a new adult dog in the same pasture with him. The resident dog is likely to see the new dog as a threat to his stock. Many livestock guard dog breeds are aggressive toward other large dogs that they do not know. Instead, allow the dogs to get used to each other through a secure fence for several days or more. Depending on their individual natures, it may take several days or even weeks for dogs to become comfortable with each other. Allow even more time for intact adult dogs of the same gender to become used to each other. In fact, intact dogs of the same gender may never get along safely alone.

Most adult livestock guard dogs will accept a young pup with no problems other than an initial "I'm the boss" moment. Moreover, there is no better teacher for a young pup than a calm, working adult livestock guard dog. A pup can be placed with an accepting adult at a very early age, as soon as you feel comfortable moving him away from his mother. The pup will model his behavior on the adult's, and the adult will often correct inappropriate actions. The pup will also be able to play a little with the adult dog, rather than trying to play with the stock.

Farm and Family Guardians

Many dogs from the livestock guard dog breeds live as general farm and family guardians, sleeping outside the home in a doghouse, attached garage, or

Several of the livestock guardian dog breeds are well suited as general farm and family guardians. And there is no better companion for you when you must do chores alone in the dark.

other outbuilding. Some livestock guard dog breeds have been selected for their ability to guard the yard and area around the house. Depending on your own situation, the presence of visitors or close neighbors, and the isolation of your property, some breeds may be better choices for this type of flexible job.

As long as he has free access to his patrol area, your livestock guard dog will easily protect your home, your outbuildings, your equipment, and your stock in and around your barn. He may not live in your pastures with your stock, but with proper socialization and good training as he grows, he will happily accompany you on your chores and enjoy spending part of his day out watching his stock. If you need to do chores alone in the dark, you can have no better companion. For many families this is a comfortable compromise, as they are free to invite their livestock guard dog inside the house for short visits. However, some livestock guard dogs are not comfortable inside and will quickly become restless in the house.

If your farm guardian does not have 24-hour access to your stock, he cannot provide total protection against predators, although if your farm is fairly compact and he has access to areas adjacent to your pastures and pens he can do a good job of warning predators away during the night. Strategically placed dog doors can give him access to areas where you want his presence, such as your garage or barn. He can also patrol areas around poultry pens and other small animal enclosures. He will often sound an alarm if he feels you need to attend to something out of place, such as stock in distress or nearby strangers. The important element is that he is available at night when predator pressure is highest, not shut up inside your house.

To raise a successful farm guardian, you need to give him the modified experiences of both a working livestock guard dog and a family companion. Have him spend some regular time penned next to stock or poultry during the period of intensive bonding. If

Protection Training

Livestock guard dogs possess instinctual responses to first warn off threats, rather than immediately attacking. They make graduated responses to threats and utilize independent judgment in evaluating potential threats. Most breeds of livestock guard dogs are unsuitable for protection training such as police or *schutzhund* work. Protection training is antithetical to their natural instincts. Encouraging aggression or forcing a "bite first" reaction is actually dangerous, since livestock guard dogs are not known to be highly obedient. A dog trained in this manner is not only a liability but also a tremendous responsibility.

In addition, there is a great danger of ruining a good livestock guard dog's temperament with the forced agitation used in protection training. Many livestock guard dog breeds have been tested by police, military, and *schutzhund* trainers who have repeatedly found them unsuitable for this type of work. When these breeds are employed by the police or military, they are generally used for patrol work with a human in a mountainous area or as a watchdog for installations.

you want his company, take him on a leash as you do your regular chores. Some folks tie the puppy's leash to their belt so that their hands are free. Praise good behavior and gently scold undesirable behavior.

At the same time, you must provide the socialization and training necessary for a family companion. However, do not bring your puppy to sleep in your house unless you want him to be a family companion rather than a farm guardian. An attached garage, a backyard, or a dog run can be an excellent solution as the home base for a young dog. He will be close enough for you to attend to him, but he will learn that his place is outside of the house unless he is invited in for a short visit. You will also have a secure place to confine him when you cannot supervise him. Remember that you will need to pay close attention to proper fencing and gates in order to keep him in the areas where he is permitted. Tying or chaining your livestock guard dog for long periods is not recommended, since it can lead to both frustration and aggression.

Family Companions

Some breeds are better suited to this role, being more amenable to the comings and goings of family and friends. Breeders can help you select a pup with a good temperament. Do not be surprised if the breeder quizzes you at length about your experience with dogs. Breeders often say a livestock guard dog is not the dog for a first-time dog owner. In addition, some breeders and breed clubs will strongly suggest that their dogs are not suitable as family pets. In any case, a livestock guard dog does not fit easily into suburban or city life; he also demands an experienced, dedicated owner.

Before considering a livestock guard dog as a family companion, you should ask yourself some basic questions. A responsible breeder will probably ask you these same questions.

Do you truly understand the basic protective nature of these dogs? Your dog will be a large and powerful canine that will seek to protect you and your family from perceived threats. He may threaten strange animals, including your neighbors' pets. When mature, he cannot be expected to calmly welcome strange visitors onto your property without a proper introduction. He might not accept strange children roughhousing with your children. Remember, your friendly pup may change drastically when he matures at 18 to 30 months.

How will you train and socialize him? Half-grown livestock guard dog pups are as exuberant as any other pups, but they weigh more than most adult dogs of other breeds. And most livestock guard dogs mature more slowly than other breeds. Are you prepared to handle a growing, sometimes willful pup that weighs 80 to 90 pounds? When fully grown, he will be a strong dog. If not well trained he will be very difficult to handle. You must be committed to socialization for the life of this dog so that he is behaves calmly around strange humans and other dogs when he encounters them on walks or as visitors to your home.

How will you keep him safe? Fencing is necessary for these dogs, and you will probably need fencing that is more robust than normal dog fencing, especially if your dog will be left alone during the day. In many urban/suburban areas electrified fencing is not allowed, so some homeowners won't be able to use a strand of electric fence to keep their dog from digging under or climbing over a backyard fence. Moreover, a livestock guard dog should not be kept on a chain or confined without lots of human interaction. Not only is it cruel to deprive a

Testing Dog Aggression

The idea of testing one dog against another is sometimes promoted as a way to determine the protective instinct of a livestock guard dog, rather than actual experience with stock. Testing in an artificial situation is not a good indicator of proper livestock behavior and may only reveal aggression toward other dogs. There is much more to a good livestock guard dog than dog aggression. Far more dogs are aggressive toward other dogs than are properly submissive to stock. Frankly, the idea of testing is sometimes used to justify dog fighting. Be very cautious of people who promote testing livestock guard dogs against each other.

flock guardian of companions, but this type of treatment may cause serious and potentially dangerous emotional or behavioral problems.

Another thing you must be aware of is the possibility of a neighbor or child teasing or provoking your dog through his fence. This activity is dangerous for both the person and your dog. Livestock guard dogs are naturally protective of their fence lines and property and may react aggressively to strangers leaning over the fence. The smaller the yard, the more protective the dog may become. You may need tall, solid fencing to protect your dog in some of these situations.

How will you provide for his exercise needs? Without exercise, you will have a very large and overactive dog in your house. He will need regular hard walks, even if he has a very large yard. He will probably not be safe off the leash unless you live somewhere extremely remote and you have done extensive recall training. Even in remote areas, he may disappear after a deer or a stray dog. Mature livestock guard dogs may not be able to handle the rude or aggressive dog behavior found at dog parks and other exercise areas. Other owners may not appreciate your huge puppy playing with their little dogs.

How will you prevent boredom? Without a job to do he will probably create one, such as chewing, digging, or barking. Speaking of barking, your livestock guard dog *will* bark. He will bark less inside a house but much more out in a yard, especially at night.

Barking, shedding, and an owner's inability to control a dog are the primary reasons livestock guard dogs kept as companion animals are turned in to rescue groups. So think carefully about whether you are capable of keeping a livestock guard dog as a family companion. It is not fair to a puppy to bring him into your home if the essential nature of these dogs is not acceptable to you.

LIVING WITH A LIVESTOCK GUARD DOG

Despite all of these warnings, livestock guard dogs can and do make excellent companions — for the right family. If the owners are committed to training, socialization, and life with a very large and intelligent guardian, a livestock guard dog is a devoted and loyal family companion. In fact, a mature, socialized, neutered livestock guard dog can be quite placid in his home as long as he receives human interaction and sufficient exercise. A well-behaved livestock guard dog is an impressive creature, and you are sure to attract much admiration and attention for your beautiful companion.

Livestock guard dogs are particularly well suited as companion dogs in rural areas, where they can receive lots of good exercise and can provide protection for their owners as well. However, livestock guard dogs also live in the largest cities and suburban areas and even work as therapy dogs. In addition to positive training, a sturdy leash and a training collar or head halter are generally essential for walking your livestock guard dog. Even if he is well socialized, you may encounter challenges from ill-trained dogs or persons he finds threatening.

Livestock guard dog pups are generally fine with the family's existing pets. When you first introduce him to the house, crate him next to the family's pets, or keep them separated for a while with pet gates. Praise tolerant behavior from your pup, and

Samsun the Kangal Dog has been well socialized to accept the attention of children.

SELECTING AND TRAINING YOUR DOG 75

use words like *easy* or *be good* when he exhibits gentle behavior. He may attempt to chase or play with a smaller pet, but you can use the same techniques used to train a young livestock guardian to shape his natural tendencies. When he matures he will probably be protective of the family's pets.

If you bring home a new pet, such as a cat, you may need to keep your dog crated or separated from

ON THE FARM

Brenda Cannon-Lelli
Beechtree Farm, Coopersville, Michigan

Located in Michigan's Lower Peninsula, Beechtree Farm is the home of a premier flock of Blue-Faced Leicesters and other native British sheep breeds such as the Clun Forest, North Country Cheviot, and Suffolk. Brenda raises sheep for show as well as stock for purebred and commercial sheep breeders. She has pursued an ambitious program to increase the genetic diversity available to her and others using imports and laparoscopic artificial insemination.

This expensive and serious program needs to be protected, primarily from coyotes, and Brenda has used livestock guardians for this purpose for 15 years. Because of her personal experience in the veterinary field, Brenda felt the most comfortable and familiar with canine behavior. She currently has four Great Pyrenees and a Maremma. The dogs are moved about the farm to different grazing fields and can work independently or in teams.

Brenda has observed that although both breeds are very loyal to their sheep, they react differently to visitors. "The Maremma is very vocal and aggressive in his posturing toward visitors to the barn and the sheep," she notes. "The Pyrs, on the other hand, will bark but then typically walk off to observe the visitors or even accept a kind pat."

Brenda breeds Great Pyrenees pups selected for the type of working dog she requires. She believes that "all or most livestock guardian dogs come by their abilities naturally from proper sire and dam selection, and in temperament testing a litter prior to placement." This selection and testing are combined with good education and exposure to livestock, so that "the dog understands its proper social order in the sheep pack."

Brenda believes that some dogs are not suited to working duty. In that case, "it is better to realize early on that the dog, your flock, and you are better off finding it a suitable 'pet' home." She strongly advises that you consider producing a litter only if you have outstanding working dogs with good health and breed traits. "Be aware that to do a proper job with a litter you will have a full-time job for four to six months, and you need to be willing to keep the puppies you produce until you find them good homes," she says.

Brenda advises new owners not to rush into their first purchase of a livestock guardian puppy: "Find out where other sheep producers have gotten their livestock guardian dogs from and if they are satisfied with the dog [and] its abilities, and if the breeder provides follow-up help and advice. . . . Don't expect the puppy to be allowed to roam the farm and come up to the house when it is cute and cuddly and then expect it as an adult to stay with the flock. And please do socialize the dog for its sake as well as your own — but do it in the pasture."

her for several days until he learns that she is a new family member. Adult livestock guard dogs will usually accept a puppy with great grace and tolerance. If the new dog is older, you should introduce him to your livestock guard dog away from your property, and keep the dogs separated for a few days until you see how both are adjusting. Many livestock guard dogs have a small dog friend that they enjoy, and sometimes the little one ends up as the boss!

Well-socialized livestock guard dogs are naturally attracted to babies and small children. Mature dogs will move carefully around them and even tolerate some poking; however, some dogs and especially adolescent pups are too boisterous to play with small children, whom they can hurt by accident. Always supervise dogs with children; even the best-behaved dog could be provoked to injure a child. Never, under any circumstances, leave strange children alone with your dog. Screaming, crying, or excitement could cause a livestock guard dog to attempt to stop the disturbance or protect his "own" child. Your dog may also protect his special sleeping spot, his toys, his food, your property, or your pets.

All dogs should be taught to respect and obey the children in your family. Involve your children in training the dog and offering treats for appropriate behaviors. Of course your children should not be allowed to tease or roughly handle your dog. Food can be an especially problematic area. Even if you have taught your dog the *trade game*, replacing one treat for another, you should never allow children to test the dog by reaching into his bowl. Never leave a dog to eat a meal or chew a bone with small children around.

Fortunately, housebreaking is particularly easy with most livestock guard dogs. Perhaps because they are such large dogs, most pups will avoid soiling their living quarters. The use of a crate for housebreaking works just as well for livestock guard dogs as for other dogs.

Solving Problem Situations

Most people who decide to employ a livestock guard dog against their predator problems will encounter one or more of these situations or problems. Fortunately, experienced users have developed several techniques and suggestions for dealing with these common dilemmas.

PROBLEM: Aggression toward Humans

SOLUTION: It is your responsibility not to place your dog in situations where he might respond inappropriately. Moreover, it is your responsibility to keep him securely at home on your property. Use signs that inform the public — **Livestock Guard Dog at Work, Do Not Disturb** — on your property and fences. Although most livestock guard dogs are capable of measured response and the ability to distinguish real from false threats, dogs have varying temperaments and may react differently from your expectations in unforeseen situations.

In addition, unfortunately, you need to protect your dog against people who don't respect your fences or signs and don't know how to behave around strange dogs. For your financial protection, make sure your dog is covered by your homeowner's insurance. The ability to show your insurance agent that your dog is not dangerous is a good reason to socialize your working livestock guard dog.

There are several situations in which you need to exercise particular caution. Livestock guard dogs are often very protective of their fence lines. Do not allow people to pet or tease your dog through his fence. Do not allow neighbor dogs or strange dogs to visit your property. Be alert if you have a female dog in heat or a new litter of puppies, since your dog may react uncharacteristically with these stimuli. Be aware that workers on your property may look like intruders carrying a weapon to your dog. Livestock guard dogs are very aware of things out of place or changes in routine. Even something as simple as a hat may make a person he has previously met look different to your dog. Roughhousing between adults or children may look like an attack to your dog. Be very cautious of strange people working with your stock in the presence of your dog. Most importantly, do not allow your dog to be alone in the company of small children.

Finally, make sure that you and your family have established a position of leadership with your dog. Breeders of livestock guard dogs often state that their breeds are not for the inexperienced dog owner. If you are unsure of your dog training skills, you and your young dog should attend classes where you can benefit from a knowledgeable trainer.

PROBLEM: Aggression between Guard Dogs

SOLUTION: Minor squabbles between dogs are normal and often sound worse that they are. However, serious or repeated conflicts need to be resolved or prevented. In most cases, two intact dogs of the same sex will not be able to get along without serious conflict. A puppy and an adult dog of the same sex will generally get along until the puppy is an adult, when there is a chance that conflict will develop. Usually this is less severe if the dogs were raised together. Also, females in heat may become aggressive toward other dogs and may cause nearby intact males to become aggressive. Keeping intact dogs obviously requires you to manage the inevitable problems.

Even neutered dogs may have disputes, however. Any two dogs may fight over food, for example. If you have multiple dogs, you should feed them separately or under supervision until you are confident of their relationship. Be especially cautious when feeding special treats.

Dogs that are upset by disturbances near their fences may redirect their aggression toward each other. The smaller the area they are kept in, the more likely dogs are to have conflicts of this type. If you temporarily remove one dog from a team, the two dogs may need to rework their relationship when they are reunited. Finally, some dogs simply do not get along with each other and will not be able to work together.

Be very careful about breaking up a serious dogfight. Never step between two dogs and grab their collars. Loud yelling or throwing water on the dogs may stop them long enough for you to remove one dog, but in most cases a single person cannot stop a furious fight. Two people may be able to pull the dogs away from each other by the hind legs, but this is not advisable for inexperienced persons.

PROBLEM: Aggression toward Herding Dogs

SOLUTION: You should securely restrain your livestock guard dog out of sight before using any strange herding dogs brought to your farm to work your stock. Your livestock guard dog will see these dogs as threats, and you don't want to punish him for reacting appropriately. If you adopt an older livestock guard dog, he may never accept you working with any herding dog, even your own, unless he was socialized to this practice as a young dog.

PROBLEM: Aggression toward Household Pets

SOLUTION: Many livestock guard dogs regard family or farm pets as animals to be protected, if they have been exposed to them at an early age or properly introduced and supervised. However, if your pet dog gets in your pasture and chases your sheep, your livestock guard dog will probably react to this perceived threat. As a general guideline, most livestock guard dogs ignore small dogs and pay more aggressive attention to larger dogs. Cats can be problematic because livestock guard dogs often see them for what they are: small predators. Quite a few livestock guard dogs distinguish between "their" cats and strange cats. The reality is that livestock guard dogs often kill small predators, and this could include any cats that stray into the pasture.

PROBLEM: Aggressive Stock

SOLUTION: Do not allow your dog to be seriously injured by aggressive stock. Although older dogs will learn to avoid aggressive individuals, it may be necessary to cull from your flock or herd an animal that is dangerous to your dog. Intact male stock animals may never accept a guard dog. They are more likely to accept livestock guard dogs if they are raised with them from an early age, although you may need to exercise caution during breeding seasons.

Alpacas and Llamas: Since llamas tend to be aggressive toward canines and are often used as guardians themselves, it seems paradoxical that livestock guard dogs are also used to protect these camelids. However, livestock guard dogs are widely used by breeders of these animals, especially as protection from packs of dogs and large predators. In most cases this relationship works very well.

From the livestock guard dog's point of view, alpacas and llamas are easily accepted as something to guard. However, if you own a camelid with a highly developed dislike of dogs, you may not be able to keep the two together. A llama, especially, could seriously injure a dog.

Additionally, it may be difficult to find reliable adult animals to leave with a pup during his socialization phase, and it is never advisable to wean young crias or to separate them from members of their group to serve as companions for a pup. You may need to raise your pup in an area next to your animals until he is large enough and old enough to be safe and trustworthy. Young dogs still need to be watched for inappropriate behaviors like chasing and fiber pulling.

Llamas and alpacas are often kept in small pastures, which may be a problem for a young or active livestock guard dog. Breeds with lower activity levels are probably a better match for small pastures with fewer animals to guard.

Cattle: Livestock guard dogs can and do work with cattle, including miniature cattle and cows with calves. Give the two species time to become acquainted through fencing. Be alert for a strong predator response by cattle toward a dog, especially when calves are present. If your cattle are pastured on sparse rangeland, they may scatter so much during grazing that more than one dog may be needed to guard them successfully.

PROBLEM: Excessive Barking

SOLUTION: Livestock guard dogs bark because barking is one of their basic tools of communication and intimidation. Working dogs bark more at night, when they detect predator threats or hear coyotes howling. Barking keeps predators away from your stock. If you cannot tolerate some nighttime barking, you should think twice about obtaining a livestock guard dog.

Puppies will begin barking early in their adolescence. You can discourage inappropriate barking by monitoring every incidence of continued barking. If the dog is barking appropriately, praise him. If you cannot detect a reason for the barking, you can scold him, but his senses are much keener than yours and you may not be able to identify the threat he has sensed. Furthermore, unless you interrupt the actual act of barking, he won't know why he is being scolded. However, if you think he is barking just to get your attention, responding will not help him stop and will actually encourage more barking.

Most young dogs bark more than adults do. At night this can be nervous barking by a young dog, since he hears so many unidentified noises. Young dogs usually grow out of nervous barking. With time and/or your continued response, he will learn which stimuli deserve an alert. Obviously, this is less of a problem if you do not have close neighbors.

Shock collars can be used to stop persistent barking, but by using them you run the risk of punishing appropriate predator warning. Never use an automatic bark control collar on a working livestock guard dog. Barking is the livestock guard dog's tool when he is working.

PROBLEM: Chasing Stock

SOLUTION: Stopping a puppy from chasing stock is part of the puppy's training. You should prevent a puppy from chasing with verbal commands or mild physical correction. Repeatedly praise good behavior around stock. If he stops his playful chasing at your verbal commands, immediately tell him *be good* or *be nice*. He will learn not to chase when you say *be good*. The pup should not be allowed unsupervised time with stock until you are comfortable with his behavior. Early mornings and evenings as well as cool weather trigger play activity in young dogs, so be especially vigilant at these times. Work off excess energy with a walk.

If your dog has a propensity for chasing, try to discover whether particular situations prompt this behavior. Is your dog chasing stock away from its food? Is the stock new and inexperienced with dogs? Have you moved the dog in with young stock? Does chasing occur only when stock are returned to their pasture? Is the chasing simple exuberance?

If your dog is younger than two and a half years or is an inexperienced older adult, patience is your best virtue. Far too many owners expect too much of a young or inexperienced dog. Give him time to mature before you give up on him. There are several methods you can use to discourage chasing; however, the first recommendation is to move back a step or two in his training. You may need to return him to "jail," as some livestock guard dog owners call the pen. Allow him out with stock only when you are walking him on a leash or supervising him. Another excellent option is to keep him with older stock that will not tolerate his attempts at chasing. In fact, many raisers of livestock guard dogs strongly recommend that adolescent dogs be kept only with older stock to avoid any potential problems.

Boredom is a primary reason for chasing and other playful behavior. Increasing the size of the area he lives in is helpful. Putting him in with another dog will provide him with acceptable social options. Some owners obtain a second livestock guard dog pup when the first dog is an adolescent. The pup may provide the company and social outlet both dogs need, but you must be very careful that the younger pup does not learn to chase from the older dog. Observe them carefully to make sure your plan is proceeding in the direction you want.

Serious chasing occurs when the dog is barking, biting, growling, and lunging at the stock. If possible, put this dog in with a dominant, experienced dog or stock animals that will teach him manners. Continued dangerous chasing is a serious matter, as opposed to exuberant chasing by a young dog. You will need to reevaluate this dog's potential as a guardian.

PROBLEM: Inappropriate Chewing

SOLUTION: All puppies need safe objects to chew. You should provide your puppy with large, fresh bones and indestructible toys. If your pup begins to chew on goats' ears, the hind legs of stock, or lambs' wool, this behavior must be stopped. Immediate verbal corrections or mild physical corrections may be effective.

You can also try a bitter-tasting product such as Bitter Apple, Boundary, Chew Guard, original Ben Gay mixed with Vaseline, Wonderdust, or another safe substance. You must reapply these products regularly. At the same time give your pup acceptable and enticing objects to chew.

If your adolescent dog begins to chew ears or wool, he is probably bored. Try verbal or physical corrections, fresh bones, and bitter products first. If he continues this behavior, you need to reassess his maturity and your ability to trust him with stock.

Intervening in Chasing Behavior

Persistent chasing can be discouraged by a shepherd's device known as the dangle stick, a light piece of wood, metal, or plastic pipe attached by a few inches of chain to a plain buckled collar (never a choke collar). But this method is only a temporary measure. Dangle sticks don't teach dogs not to chase, they simply punish them if they run.ABangle sticks can damage the bones and joints of a growing dog if they are not used carefully.

You may receive suggestions about attaching a long chain or a chain attached to a tire or log to your dog to prevent chasing. This is called a drag, and it slows a dog down. A drag should not be left on a dog for long periods. Again, it does not teach a dog not to chase. The drag chain can be from 3 to 10 feet in length and is always attached to a plain buckled collar, not a choke chain. If your dog is escaping through fences or you fear he will become entangled, use a longer chain so that the dog is not strangled.

NOTE: Both danglers and drags can be dangerous in a barn or anywhere they might become hooked on something. Check often on a dog if you are using one of these devices, since a dog can dehydrate rapidly in hot weather if he becomes tangled in something away from water. Be aware that the drag may also prevent your dog from responding to a predator.

Some owners resort to tying up the dog or using a "zip line" in a stock enclosure. If you cannot monitor your young dog any other way, this may be your only option. An overhead zip line is safer than tying up the dog, for both the dog and the stock. Again, this should be a temporary measure, not a permanent arrangement. Always use a plain buckled collar if you must tie your dog.

Shock collars are controversial and should be considered a final option to stop unwanted behavior. The electric shock must be given exactly when the dog is behaving inappropriately. It should be used to solve only one problem at a time or the dog will be confused. If you consider this option, please learn the correct methods for using the collar. Consider working with a professional trainer if the situation has reached this point. Be cautious about using this collar in the presence of another dog or a person, since some dogs react with redirected aggression when they receive the shock. Do not use shock collars on young puppies.

If he can't be moved to a larger area or moved in with older stock that will not tolerate this behavior, you need to step back in his training and provide increased supervision.

PROBLEM: Dead or Injured Stock in the Pasture

SOLUTION: Do not assume that your dog has killed a dead stock animal you find in your pasture unless you have firm evidence of it. If the carcass is that of a newborn, it may have been stillborn. Its mother may also have abandoned it, perhaps because your young dog unwittingly interfered in a birth or with the newborn — which is why he requires your attention in training him to accept the birthing process. A young or distracted dog may occasionally lose an animal to a predator, or he may be overwhelmed by multiple predators. He may chase off a predator after a serious injury to stock. An animal may die from natural causes.

In any of these cases a dog may innocently investigate a dead animal or begin to eat it, confusing your ability to sort things out. As a natural instinct, canines often eat their own dead offspring to discourage predators. See chapter 1 for signs that can help you determine the cause of death or injury.

It is possible, however, that your young dog will injure one of your animals despite your best efforts at training and monitoring play behavior. This is not a reason to give up on your livestock guard dog. Some experienced livestock guard dog users say that there is one dead or injured animal in every livestock guard dog's life. This is certainly not true for every dog, but it is common.

If he is young, you need to redouble your efforts. You cannot expect him to reach reliability until maturity, which can vary between 18 and 30 months of age. Raising a livestock guard dog is a long-term project, and you may suffer setbacks. It is unfortunate that many dogs are turned in to rescue organizations at this point, when they are still salvageable. It is very rare for a well-bred, well-trained dog to actually turn into a confirmed stock killer.

Obviously, you should remove all dead animals from your pastures as soon as possible. Carcasses attract other predators, breed flies, and spread disease, and they could encourage inappropriate resource guarding by your dog.

PROBLEM: Digging Dens

SOLUTION: Many adult livestock guard dogs dig dens in which they can retreat from high temperatures or cold winds. This is an entirely natural behavior. If these dens bother you or are a hazard, make sure that you have provided your dog with a cool or sheltered place for a retreat. Some owners report that placing dog feces in the den discourages its further use.

PROBLEM: Kenneling Your Dog

SOLUTION: At times it may be necessary to kennel your livestock guard dog, such as when he is recovering from an injury or surgery or when you are temporarily keeping your stock in a small feedlot or birthing sheds. Most dogs have no problems with short periods of kenneling and are eager to resume their life with their stock. Do make sure your dog receives regular exercise while he is kenneled. Your dog will appreciate some free time in nearby paddocks or pastures, even if his stock is not there. Some dogs become visibly depressed and miserable when they are prevented from being with their stock. If so, try to minimize this situation or find a few animals he can stay with.

PROBLEM: Fear of Thunderstorms

SOLUTION: Some dogs are frightened of thunder or other sudden loud noises. Otherwise reliable livestock guard dogs may leave their stock during these times to seek shelter or to be closer to you. Punishment is not appropriate or helpful in this circumstance. Fortunately, predators are not usually active during thunderstorms. As long as the dog remains in his enclosure and has a safe retreat, his fear and behavior are not a serious problem. However, a few dogs become extremely distressed and may injure themselves in this situation. These dogs should be placed in a safe location if possible.

A dog with a serious, self-destructive thunderstorm phobia may not be suitable as a full-time livestock guardian; he would function better as a farm guardian with access to a barn or garage. Some owners have found relief for their dogs with herbal remedies or medication prescribed by a veterinarian. Behavioral modification techniques that are often prescribed for thunderstorm phobia are usually not practical for a working dog. You can avoid conditioning a dog to fear thunder, gunshots, or other loud noises by ignoring them in his presence. Do not comfort a fearful dog, as this reinforces fearful behavior.

Is My Dog Really Working?

You might think your dog is not really working, especially since you see him lying around all day. It is common for livestock guard dogs to sleep through the heat of the day, when predator pressure is low. Sure, you hear some barking at night, but you haven't seen any predators. In truth your dog may be doing such a good job that the predators are visiting other farms. Never fear, 95 percent of livestock guard dogs *are* aggressive toward predators.

If you are losing some stock despite the best efforts of your dog, you may need a second dog. Your stock may be spreading out too much in a large pasture for one dog to protect successfully, or your predator pressure may be too intense for one dog to handle.

If your dog suddenly stops doing his job, ask yourself these questions. If he is intact, does a nearby female in heat distract him? Is he feeling well? Are some of your stock animals behaving aggressively toward him? Good working dogs don't just stop working. Find the reason behind the behavior, and then do what is necessary to correct it.

5
TAKING CARE OF YOUR DOG

A hardworking livestock guard dog needs access to clean water, good food, and shelter from the heat of the sun, driving rain, and snowstorms. Your livestock guard dog also needs routine health care: immunizations; annual heartworm blood tests and preventives; regular worming and protection from flies, fleas, and ticks; and nail trimming and grooming. There are many excellent books to guide you in caring for your dog, but here are some special concerns for working livestock guard dogs:

- What food is appropriate for a giant-size dog?
- How will you keep his food safe from the stock?
- How will you arrange for veterinarian visits?
- How will you prevent cross-contamination of parasites between your dog and stock?
- How will you keep up with grooming a very large, hairy, outdoor dog?

Feeding Your Dog

What to feed dogs has become quite controversial. Different experts advocate raw-food diets, all-natural diets, or high- or low-protein kibble. In their native countries these dogs are still raised on low-protein, low-fat diets, but this diet often affects their size and health negatively. In North America, on the other hand, many dogs are overfed. Some owners report that feeding lower-protein food reduces the activity level of an adolescent or adult dog. Others insist that the key to their dogs' health is high-quality protein sources (such as raw meat and bones) coupled with low-carbohydrate calories.

Do your homework and experiment to discover what works best for your dog. Some owners find a compromise by feeding a high-quality kibble supplemented by raw meaty bones, such as skinned chicken necks or frames, and other healthful foods.

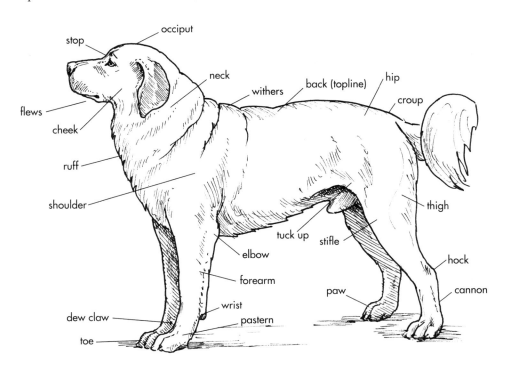

Whatever your personal decision, a working dog deserves sufficient good-quality food that he can eat without competition or aggression from other dogs or livestock. Do not listen to people who say he can hunt for his own food.

FEEDING PUPPIES

Most livestock guard dog breeders feed their puppies a good-quality puppy food. Most do not feed puppy food for an entire year, as some manufacturers recommend, instead switching the pups over to adult food at three to four months of age. Many breeders suggest limiting protein to 23 percent or less, although the optimal protein level continues to be debated.

What is more important for growing puppies is to limit caloric intake. Puppies need to be kept slim but not thin. You should be able to easily feel their ribs and see a tuck-up in the loin area. The same is true for adult dogs. Overfeeding large-breed puppies is a serious concern; research has recently shown a strong correlation between excess caloric intake and the onset and severity of hip dysplasia. Other orthopedic problems can also be worsened by poor diet and excess body weight in growing pups. Many livestock guard dog owners feed their older pups a good-quality adult kibble formulated especially for large breeds. These products are lower in protein and calories than other brands of puppy food.

> **Feeding Raw Food**
>
> If you choose to feed raw meat or fresh bones to your dog, you need to find sources of good-quality, fresh ingredients and educate yourself about the proper handling of raw products. There is some danger of splintering or choking when dogs are allowed to chew any bones. In general, fresh raw bones are always safer than cooked bones.

Feeding a quality food is more important than giving inexpensive food in quantity. Good-quality food costs more, but the dog needs to eat less of it. Meat should be the first listed ingredient. Some breeders advocate avoiding corn, which is often linked to food or skin allergies, and others advocate avoiding wheat as well, for the same reasons. Inexpensive food with low-quality ingredients can be a very poor bargain because it may reduce your dog's health and ability to work.

Feed a puppy only what he will eat in ten minutes, twice a day, supplemented with fresh, raw bones or other safe items for chewing. Do not overstimulate him with praise or petting while or just after he eats, since some dogs can become conditioned to eat only in your presence or with your attention.

FEEDING ADULT DOGS

Adult dogs may be fed once or twice a day. Some owners divide the feeding into two portions to reduce the likelihood of bloat. Feeding times are good opportunities to check your dog's condition and give him some attention. You may need to increase your dog's regular ration during winter or times of increased predator pressure, when he is expending more energy. If your dog is overactive, you can try reducing his protein or caloric intake. Obviously, a pregnant or nursing bitch needs particular attention paid to her diet.

Having your dog self-feed is less desirable than feeding him yourself, as it reduces your chances to monitor your dog and because the food is exposed to weather and insects. Your dog may also overeat. However, self-feeders are sometimes necessary in large farm, pasture, or range conditions. In this case the feeder should be secured away from the livestock. Many owners convert the puppy pen or a similar area into a feeding and retreat area for their livestock guard dog, giving him a porthole or other

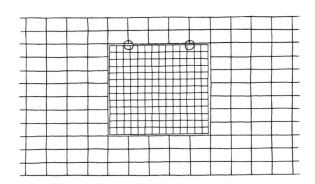

A dog door or porthole in a fence allows a dog access between pastures or to a feeding station.

A jump feeder is another alternative for a dog feeding station in a pasture.

entrance inaccessible to the stock. When you first begin to use a self-feeder, monitor your dog carefully to be sure he is actually eating. Dogs do not naturally know how to use these feeders and some dogs do not like them at all.

Dogs may aggressively protect their food, special treats, or bones against stock. In this case, you may need to situate your dog's food in a place where he feels more secure while eating. Also, try feeding your stock and dog at the same time, which keeps them both busy.

Adult dogs may stop eating at times due to hot weather, hormonal changes in an intact female, the presence of a female in heat, or sometimes for no clearly identifiable reason. Always be alert for signs of illness, injury, or poisoning if the dog continues to refuse food for more than a day or so.

Almost all owners of livestock guard dogs report that their dog never eats the amount of food recommended on the bags of commercial dog food. These recommended amounts seem to be based either on very high-energy dogs or on a desire by the manufacturer to sell more food. Most livestock guard dogs have low metabolisms, and, especially as adults, they are far more observant than active.

NOTE: Problems may arise when multiple dogs are being fed, especially if there are dominance issues among the dogs. If you cannot feed multiple dogs without problems developing, they may need to be placed in separate locations while they eat. You may not be able to feed special treats if they cause problems with multiple dogs.

PROVIDING WATER

If you rely on stock tanks to provide water for both your dog and your stock, be certain that your livestock guard dog can reach the water level and that the stock do not interfere with the dog obtaining water. If you use automatic waterers or livestock fountains, make sure your dog can use them. In areas of freezing temperatures, dogs need access to heated water. Working dogs should never be forced to survive on snow or ice.

Make sure your dog can reach the water level in your stock tank. Some dogs will also use stock tanks for cooling off!

Shelter for Your Dog

Although livestock guard dogs are equipped for outdoor living in difficult conditions, they do need shelter from heat, cold, rain, sleet, and snow. It is unfair to expect a working dog to suffer through any of these conditions without shelter. High temperatures and humidity can cause heatstroke in hardworking dogs. Heavy-coated dogs are particularly vulnerable. In very hot weather you might find your livestock guard dog standing in your livestock tank, burrowed in a cool den, or in another cool spot during the day.

On the opposite end of the spectrum, although quite often you will find your livestock guard dog out lying in the snow enjoying the cooler temperatures of winter, he does need a spot in which to shelter during times of severe cold. During inclement weather some dogs enjoy curling up in the barn or shed with their stock. Others will use a good-size doghouse, although they may find themselves with a goat or sheep as a visitor. If your dog is nervous about sharing his doghouse with stock, place the shelter with his feeding station in a stock-proof enclosure in the pasture. Note that providing a doghouse for your dog will not prevent him from guarding his stock.

Proper Health Care

Livestock guard dog breeds are very stoic. You need to pay careful attention to notice lameness or illness in some individuals. Changes in behavior, a dull or ragged coat, a lack of appetite, weight loss or gain, unusual thirst, and difficulty in lying down or getting up are all signs of poor health or pain.

When it is time for the dog to see a veterinarian, some owners prefer to have the work done at home as part of a farm call. If you arrange it ahead of time, many farm vets will perform annual heartworm checks and vaccinations for your dog when they come to treat your stock.

When they must bring a dog to the vet's office, some owners prefer to keep the dog in their car or truck at the clinic parking lot, asking the vet to come out to the car to examine the dog. The most preferable method is to accustom your dog to actually visiting the vet's office while he is a puppy. Most livestock guard dogs are not territorial or aggressive away from home and behave quite well. The slippery floors may be your greatest challenge!

Some guard dogs will sleep with their stock in a shed, while others prefer their own doghouse.

ANNUAL VACCINATIONS

Livestock guard dogs need the same basic vaccinations as any other dog, including those for adenovirus, parvovirus, distemper, and rabies. Additional vaccines that you might consider, depending on conditions in your region, include bordetella, coronavirus, giardia, leptospirosis, and parainfluenza virus. A vaccine against Lyme disease is also available, though unless you live in an area of very high tick infestation, flea and tick preventives are generally enough protection against this disease.

Puppies require a series of vaccinations during the time when their maternal antibodies are diminishing. The recommended ages for puppy shots and the required frequency of booster shots for adults are subjects of debate, and different areas of the country have increased incidences of specific infectious diseases. Your own veterinarian is the best source of information for your dog and your area. Requirements for rabies vaccinations vary by state and the inoculation must be given by a veterinarian to be legal in the event of a biting incident. You may choose to give other vaccines yourself. Keep careful records as a future reference.

DEALING WITH PARASITES

Among the most common health problems livestock guard dogs experience are internal and external parasites, many of which affect both the dog and the stock he guards. You certainly want to minimize cross-contamination. Scratching, licking, and biting are signs the dog is troubled by something.

Tapeworms are a special concern for livestock raisers. Tapeworms can contaminate sheep carcasses and pose a threat to human as well as animal health. Two tapeworms associated with dogs, *Taenia hydatigena* and *Taenia ovis*, can cause lamb carcasses to be condemned at slaughter. All farm dogs should be regularly tested for tapeworms and treated as needed.

Parasite Control

A wide range of parasite remedies is available. However, the flea, tick, mite, and fly remedies used on livestock are not always safe for dogs. Your veterinarian can suggest good flea and tick preparations for canine use. Some owners prefer herbal remedies, often using garlic to discourage external pests. Biting flies, in particular, can make an outdoor dog's life miserable and cause infection. You can discourage biting flies with pyrethrin or permethrin sprays or ointments, and you can use diaper rash or fly repellent ointments for increased protection on wounds.

Fleas can be controlled with topical or oral preparations, some of which also repel ticks, mosquitoes, and lice. Generally outdoor dogs require regular and robust protection from these parasites, depending on the area of the country and the climate.

Dogs can acquire roundworms, hookworms, and whipworms from contaminated soil or feces. Over-the-counter and prescription medications are available for the treatment of these internal worms.

Heartworms are now found throughout North America. The larvae are transmitted between animals through the bite of mosquitoes. Several medications are available for prevention of this parasite; most of these medications also control other parasites as well. Some owners use livestock

Parasite Medications

Medication	Parasites Controlled
Ivermectin (Heartgard, Iverhart, Trihart)	heartworm, roundworm, hookworm
Selamectin (Revolution)	heartworm, roundworm, hookworm, fleas, ear mites, mange mites, ticks *(Dermacentor variabilis)*
Milbemycin oxime (Interceptor)	heartworm, roundworm, hookworm, whipworm
Milbemycin oxime-leufenuron (Sentinel)	heartworm, roundworm, hookworm, whipworm, fleas

worming preparations such as ivermectin as a heartworm preventive for their dogs, but you should consult your veterinarian before attempting this, since there is a potential for serious side effects.

ROUTINE CHECKS

You can prevent many veterinary visits by taking good care of your dogs yourself and regularly inspecting them for parasites, wounds, itchy spots, and other problems. Here are some things you should routinely look for as you work with your dogs.

Care of Teeth

Fresh, meaty bones, dry kibble and biscuits, and rope or hard rubber toys can all help keep teeth clean. Signs of serious dental problems include bad breath, lack of appetite, excessive drooling, and pawing at the mouth.

Care of Ears

Pay attention to head shaking and pawing at the ears. Ear problems can be caused by mites but also by moisture, allergies, or bacteria. Humid conditions can encourage fungal and bacterial infections. Cleaning the ears, trimming excess ear hair, and using drying agents can help prevent ear problems. Outdoor dogs can easily get burrs, foxtails, and seeds lodged in an ear. Check your dog's ears daily if this is a problem. You can remove visible objects with blunt tweezers.

Care of Eyes

Outdoor dogs can get seeds or other debris in their eyes, especially in tall grass. If you see tears, squinting, blinking, or pawing at an eye, flush it out with a sterile saline solution. An untreated eye injury can result in a corneal ulcer. If the problem continues or the eye develops a discharge, see a veterinarian to remove the object and determine whether an antibiotic is required.

Care of Feet

Hardworking dogs need regular foot inspections. Although your dog may naturally wear down his nails on hard ground, his nails will probably need regular trimming if he is primarily on soft ground. Keep the toenails well trimmed, so that they barely touch the ground. Long toenails can cause the foot to splay out. Pay attention to any dewclaws, since they can grow into the pad itself.

If your dog is limping, check for injuries to his pads or toes, looking for cuts, punctures, and thorns or similar objects. Clean punctures and cuts with an antiseptic. If the wound is infected your dog may need a tetanus shot; if it is deep the wound may require stitches.

SKIN PROBLEMS

Food and environmental allergies and fleabites can cause dermatitis, a condition characterized by severely itchy, inflamed areas. Antihistamines and corticosteroids can offer relief, but flea preventives, a strict elimination diet, or environmental allergy testing are steps toward long-term solutions. If left untreated these dermatitis patches can become hot spots. Hot spots can also develop from a lack of grooming or even stress. They are more likely in hot and humid weather, especially in the dewlap area under the neck, at the collar.

At first hot spots look swollen, hot, and irritated, but they rapidly progress into open sores that ooze pus. Left untreated, they can eventually cover large areas and make your dog vulnerable to serious infection. A hot spot should be clipped free of hair, cleaned, and covered with a topical antibiotic. Some owners treat small hot spots with medicated powders or even cornstarch, which helps dry the area. Large hot spots may require antibiotics or corticosteroids.

Your dog licking and chewing at an irritated area can lead to the development of a granuloma, or a thickened area of red, hairless skin. Granulomas are often found on the front legs, where the dog can easily worry at them. You may need to apply a bitter substance to the area to discourage the dog from licking at it until it heals.

Mange, caused most often by the *Demodex* mite, is not contagious. Mange appears as scaly red skin with hair loss, usually on the face or front legs. Small or localized areas of mange can be treated with an ointment to kill the mites. Widespread or generalized mange is usually hereditary and difficult to treat. In this case the dog will need to be dipped in a mite-killing solution once or twice a week, and antibiotics and specific shampoos may be necessary. Do not breed dogs with generalized mange.

Scabies, characterized by itchy, mangy spots, is caused by the *Sarcoptes* mite. Scabies is very irritating to dogs; biting and scratching these spots leads to crusty sores, hair loss, and even infection. Scabies is contagious for dogs, cats, and people. You will need specialized treatment for both the infected dog and any dogs that have come in contact with him.

WHAT ABOUT GROOMING?

Don't expect your working livestock guard dog to look like a show dog! Many livestock guard dogs are bathed rarely, if at all. Overwashing the coat of a dog that lives outside removes the protective oils that provide some weatherproofing. Mud will generally fall off once it has dried.

The longhaired breeds do require some regular coat care. A heavy coat can provide protection from both weather and potential predators, but it can also collect an amazing amount of debris and tangle into uncomfortable mats. Some owners give their dog a thorough grooming once a year, in late spring or early summer, clipping heavily matted areas (such as the underside, the area under the tail, and the feathering on legs and feet), removing dead hair and the shedding inner coat, and giving the dog a good bath. Many owners feel that the best approach is

The corded coat of a Komondor requires special attention and care, which can be learned from someone experienced with the breed.

a shorter monthly grooming session, with careful attention paid to the areas that form mats: behind the ears, around the neck, under the legs, under the tail, and in the rear. A conditioner will soften mats so they can be combed out.

Breeds like the Puli and Komodor that form cords in their coats are special cases, and you should consult your breeder for assistance in grooming decisions about them.

THE OLDER DOG

Life span varies among breeds, but most livestock guard dogs have active lives of 10 to 12 years or more. Some livestock guard dogs will work well into their old age with no signs of a slowdown until they near the end of life. Others will become slow and stiff or lose some hearing and vision but still be capable of astonishing action when they perceive a threat. In their native homelands, aged livestock guard dogs were often retired to a more comfortable place in the yard or barn, where they could still participate in farm life. After a lifetime of hard work, some livestock guard dogs appreciate a well-earned retirement. A few make the transition to a beloved house pet. Other dogs strongly prefer to be out and among their charges, although if predator pressure is constant, the older dog may struggle to be as effective as he once was. The best solution may be to provide him with a small area to patrol and a smaller number of animals to watch, along with a comfortable place to rest.

Before your working dog is too old to guard well, you should consider raising another pup to take his place. The time to think about this is when your working dog is about seven years old. The pup needs time to grow and learn, and clearly there is no better teacher than an experienced working dog. Most livestock guard dogs will accept a nonthreatening pup, which will slowly step into his mentor's footprints.

Specific Health Concerns

Larger breeds and some individual livestock guard breeds have particular health issues that you should be aware of. Though they are rare, it is important that you be cognizant of potentially serious problems.

BLOAT

Bloat is a condition in which the stomach becomes overstretched and may twist completely, causing death if not treated promptly. Some breeds are more prone to bloat than others. This condition is more common in deep-chested dogs, usually after they eat rapidly and/or drink large amounts of water, often immediately before or after exercise.

The initial signs of bloat include restless pacing or other signs of discomfort or pain, gagging, unsuccessful vomiting, and a swollen abdomen. Eventually the pulse weakens, breathing becomes shallow, and the dog is unable to move. Even in its early stages bloat is an emergency situation. A veterinarian may be able to relieve bloat with a stomach tube, although X-rays and surgery may be necessary if the stomach has twisted.

To prevent bloat, follow these precautions:

- Feed two or three smaller meals daily, rather than one large meal.
- Restrict water and exercise for an hour after eating.
- Do not allow your dog to drink a large amount of water after exercise or a stressful experience.

EYELID CONDITIONS

Ectropion is a condition in which both eyelids turn outward. It is more common in some breeds and therefore is thought to be genetic. In some cases, the condition develops or becomes more troublesome as the dog ages. It is correctable by surgery. Entropion is a condition in which the eyelids turn inward. Like ectropion, it is more common in some breeds. It is detectable in puppies and is also correctable by surgery after four to six months of age.

BONE AND JOINT ISSUES

Large dogs are vulnerable to bone and joint injuries. However, you can reduce the risk of your dog incurring such injuries by taking the following precautions:

- Prevent your dog from exercising too hard, jogging long distances, jumping off high objects, and walking on slippery surfaces for his first year or so. Even though he may be the size of a full grown dog, remember that he is still a puppy.
- Do not overfeed puppies or grown dogs.
- Make sure your puppy is fit for strenuous activities. Full-time livestock guard dog puppies generally receive enough constant exercise to be fit for their jobs.

ACL Injuries

A ruptured cruciate ligament in the knee is commonly referred to as an ACL (anterior cruciate ligament) injury. This injury most commonly occurs when the dog twists on his hind leg, as may happen if the dog slips on a slick surface or makes a sudden turn while running. Poor conformation, excessive weight, and lack of regular exercise may increase the chances of this injury.

The dog may appear suddenly lame and hold his foot off the ground. While this injury may appear to get better on its own, the lameness usually returns. A partial tear of the ligament may heal with rest and anti-inflammatory medications. A full tear in a large dog usually requires surgery for repair. Different surgeries are available, but they all require a period of rest and rehabilitation. Strain on the opposite leg may cause ligament injuries in that joint as well.

An ACL injury may cause arthritis later in life. Dietary supplements such as glucosamine, chondroitin, methylsulfonylmethane (MSM, an organic sulfur compound), and vitamin C may prevent or relieve this arthritis.

Panosteitis

Sudden and severe lameness or "growing pains" in young dogs is known as panosteitis. "Pano" tends to begin in one or both of the front legs in dogs from five to fourteen months of age, and is more common in males than in females. The pain may move from leg to leg and the dog may have periods of relief followed by a resurgence.

The exact cause is unknown, although some breeds, particularly large ones, seem more prone to it than others. The condition will end on its own, although the pain can be relieved somewhat by a period of restricted activity and pain relief medication.

Osteochondrosis (OCD)

OCD is a disease of joint cartilage that is often found in rapidly growing large-breed dogs. OCD can develop in the shoulder, elbow, hock, and knee. It is not believed to be genetic, although there may be a tendency toward OCD in some family lines. Controlled feeding to limit the rate of growth may prevent OCD. The condition is usually treated with rest, although sometimes surgery is required.

Osteoarthritis

Just like humans, aging dogs can develop arthritis. Dogs usually develop arthritis in the hips and elbows. Arthritis can be relieved with doses of glucosamine, chondroitin, MSM, and vitamin C. Fish oil, partnered with vitamin E, may be an anti-inflammatory as well. Human doses of these supplements can be cheaper than preparations made for dogs. Aspirin and other pain relievers may be necessary as your dog ages.

Hip Dysplasia

This abnormality of the hip joint may lead to inflammation, arthritis, lameness, and pain. Genetics plays an important role in the development of this condition, but overfeeding of large-breed puppies and young dogs may also aggravate any potential problem.

Signs of hip dysplasia include limping and reluctance to move freely. Although severe hip dysplasia may require surgery, mild cases can be made more tolerable with the use of glucosamine, chondroitin, MSM, and vitamin C. Again, human doses of these remedies can be cheaper than preparations made for dogs.

Keeping your dog lean and active with good muscling will also help relieve symptoms. Mildly dysplastic dogs are often able to work with no signs of impairment for several years. Some cases may require anti-inflammatory or pain medication.

> ## Skunked!
>
> **If your dog lives outdoors** and *you* don't mind, the smell will wear off in a couple of weeks. If he has been sprayed in the face or eyes, he does need to have his eyes flushed with clear water or human eyewash. Skunk spray is not blinding, but it is very painful to the eyes. Your veterinarian can recommend soothing eye drops that you can give to your dog for a few days.
>
> The commercial deskunking solutions are good, but you probably have the ingredients for an excellent homemade solution in your house. The amounts given below make enough solution to completely wash a medium-size dog. You can add a quart of lukewarm water to dilute the solution for a larger dog, but a dog with a thick double coat may need two or three times this amount.
>
> - 1 quart unopened, 3 percent hydrogen peroxide
> - ¼ cup baking soda
> - 1 or 2 teaspoons liquid soap (Liquid Ivory or Softsoap is better than dishwashing liquid or shampoo)
>
> Mix the three ingredients in a plastic bucket; they will fizz. The solution loses potency rather quickly, so use it right away. Wear rubber gloves to protect your hands (the solution stings open cuts), and wear old clothes. Work the solution in thoroughly, and let it remain on the dog for about five minutes. Then rinse thoroughly. You may need to repeat the whole treatment.
>
> **CAUTION:** This solution may lighten your dog's hair.
>
> If you are concerned about using hydrogen peroxide, you can also try a mixture of two-thirds water and one-third apple cider.

Livestock guard dogs that are to be used for breeding should be screened for signs of hip dysplasia through the Orthopedic Foundation for Animals (OFA) or the University of Pennsylvania Hip Improvement Program (PennHIP).

Elbow Dysplasia

Elbow dysplasia is a general term for an abnormality or degenerative problem of the elbow joint, sometimes related to OCD. It can also occur when bone fragments are present in the joint, possibly from an injury. Excessive weight, overuse, or housing on a hard surface can also stress the joint, especially in a giant-breed puppy. Elbow dysplasia is far less common that hip dysplasia in most livestock guard dog breeds. In those breeds for which is does tend to be a problem, screening for the condition is available through OFA.

Osteosarcoma

This is the most common bone cancer found in dogs, generally occurring in large to giant-size breeds, most often in the leg bones. Limping is the earliest sign of a potential problem. Diagnosis is made through X-ray and/or biopsy.

Thinking of Breeding?

Please think seriously before breeding your dog. Keeping your dog intact creates more problems for you throughout the dog's lifetime. Intact dogs are less attentive to their jobs at times, as they are more likely to be distracted by females in heat and often are more aggressive. At times you will need to separate your intact dogs or prevent your intact female from being bred by neighborhood male dogs. Also, consider the time your female dog will not be able to work if you breed her. Do not undertake breeding unless:

- You have become thoroughly knowledgeable about your breed, including behavioral, health, genetic, and pedigree issues.
- You have screened your dog for all potential health concerns in his breed.
- You have determined that you will be able to place all the puppies in good homes. Most livestock guard dogs have large litters of 8 to 10 puppies.
- You accept that you may actually lose money rather than making it from selling puppies. Treating unexpected medical issues in the mother or puppies, feeding and health-screening the pups, and taking care of their basic medical needs may cost you more than you anticipate.
- You have the time and space to socialize the puppies to stock and people. The puppies will also need safe housing and fencing.
- You are prepared to advertise your pups, talk to potential buyers on the phone or by e-mail, welcome potential buyers to your farm, and work with the buyers to transport the pups.
- You have the knowledge to teach new owners about livestock guard dogs. This also means you have successfully raised dogs all the way to adulthood — not just one dog to age two.
- You are committed to helping the new owners and to taking back your pups if necessary.
- You are prepared to deal with unhappy owners and owners who renege on their contracts with you.

Breeding a litter of puppies is an emotional experience. After investing so much time, money, and love, you want to feel good about the situations your lovely little pups will live in.

Quilled!

There are several myths about porcupine quills. First, porcupines do not throw or shoot their quills, but they do whip their tails swiftly, leaving large numbers of quills in anything the tail touches. The quills are hollow, but you do not have to cut them to relieve the air pressure before you can remove them. Tiny pieces of quill that are embedded in a dog do not need to be removed, especially if the dog receives an appropriate antibiotic. Finally, the quills do not actually have barbs, but they do have tiny scales tapering away from the point, making them somewhat difficult to remove.

If your dog is a victim of a porcupine encounter, there is no reason to panic, although your dog may be frantic and in pain. After their first fruitless attempts to rid themselves of the quills, most dogs will calm down. If there are only a few quills and you are certain there are none are in your dog's eyes, mouth, tongue, or throat, you may be able to remove them yourself. Cover the dog's eyes with your hand or a cloth, and then firmly grasp the quill near the base with pliers and pull. The dog will often jump backward and pull the quill out himself. You may need to let him settle down between attempts.

If your dog has a large number of quills or if there are quills in a potentially dangerous area, such as the throat, he needs to be taken to a veterinarian, who can administer anesthesia and carefully remove all of the quills. Left untreated, quills can cause infections and painful sores.

6 Livestock Guard Dog Breeds

The dogs used to guard livestock have been developed and bred throughout a wide sweep of southern Europe and central Asia. These dogs have many similarities in behavior, and at times they often look quite like one another. In many cases, efforts in their native countries to recognize them as separate breeds are quite recent. Some may argue that there are few differences between them and that they can be interbred freely.

However, national and cultural pride is a reality. The Food and Agricultural Organization of the United Nations has long recognized that a breed is, in practicality, whatever the people in an area regard as a breed. As another complication, the Fédération Cynologique Internationale (FCI) recognizes many landrace breeds by assigning the ownership of a breed to a specific country, even though breed development does not always respect national borders.

Despite these issues, and although livestock guard dogs are related in function and appearance, we must recognize that each group of people in a different area has made conscious choices with its dogs, selecting them for traits specifically adaptable to that group's particular geography and agricultural needs. We are certainly learning more about the real, functional specializations between breeds that may share a distant common ancestry and therefore a very similar phenotype or appearance. Individual livestock guard dog breeds have become specialized for different kinds of work that require different combinations of behaviors. These behavioral differences should be treasured, because they increase our ability to choose the right breed for our specific situation. If we accept that the people in the small geographic area of the United Kingdom could develop some ninety breeds of sheep, each suited to a specific geography and human need, then surely we can accept numerous livestock guard dog breeds. Just as there are a multiplicity of terriers and gun dogs, the reality is that there are many different breeds of livestock guard dogs.

	Breed Vocabulary
BREED	Any group of animals that is reproductively isolated through either geography or human manipulation. The members of a breed resemble each other and produce offspring that also resemble them. A breed has common behaviors or abilities. Since the nineteenth century, the term *breed* has come to mean the same thing as *standardized breed*, or a group of animals with a standard, pedigree information, and a formal registry.
LANDRACE	A native breed without a formal registry, although its breeders may keep informal pedigrees of their animals. Landrace breeds generally have greater diversity of appearance than standardized breeds.
PUREBRED	The offspring of registered or pedigreed animals.
PEDIGREE	A record of ancestry.
TYPE	The conformation, character, and behaviors that make a group of animals distinct from another. A breed may have an idealized type or several acceptable types.
PHENOTYPE	The exterior appearance and observable behavior of an individual. Phenotype cannot be used exclusively to determine breed.
GENOTYPE	The genetic package of an individual.

Developing Standardized Breeds

Many livestock guard dogs breeds are making the journey from being landrace breeds without a standard or registry to being recognized breeds with standards, pedigrees, registries, and organizing clubs of breeders. Some breeds, such as the Great Pyrenees, have already completed the process, while many others have barely set foot upon the path. Westerners should guard against any tendency to sweep into the native homes of these dogs and make pronouncements about which dogs are breeds or what names they should have or what standards they should be bred to. As recent adopters and users of livestock guard dogs, we have much to learn from the people who have actually used these dogs for hundreds of years, and we should listen carefully and respectfully to their advice and opinions.

In addition, these native users of livestock guard dogs will continue to make their choices based on their own evolving situations, which may include the increased presence of large predators, predator conservation efforts that discourage or forbid lethal controls, or the decreased need for sheep- or goat-guarding dogs as urbanization and other development occurs. This is an exciting and sometimes turbulent time for emerging breeds and national efforts to promote them. The situation is certainly somewhat fluid, as fanciers within each country sort the paths their breeds will take. Even the breed names can be under dispute; in some cases breeds have been known by Westernized names rather than their original native ones.

In the future, as the process of breed development proceeds, we will probably see more clearly defined differences between these breeds. And of course we will make our own decisions here in North America as well. The most important thing that we must guard against is breeding solely for show or companion homes, rather than breeding to preserve the basic working nature of these dogs. We must appreciate and foster the differences among these breeds so that the breeds remain distinct, thereby offering real choices for different situations that would benefit from different types of dogs.

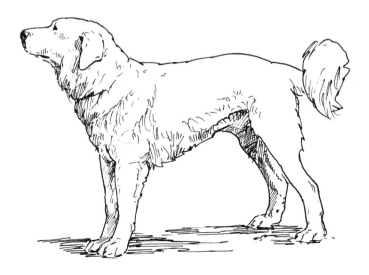

Landrace breeds, such as the Akbash, exhibit differences or diversity in type. Individual dogs may be heavier or lighter in build or have longer or shorter coats.

One interesting feature of landrace breeds is the diversity of type they manifest, as opposed to the uniformity of some long-established Western breeds. This diversity reflects a broader genetic base, which is valuable in a rare breed with a small population. Rare breeds sometimes require new dogs to be imported into the gene pool, even though they lack pedigrees or registrations like those found in the American Kennel Club (AKC) breeds. Many of the rare livestock guard dog breeds have independent registries that allow these imports or use the services of the United Kennel Club (UKC), which allows for the single registration of approved dogs. Since the AKC allows only dogs with three-generation pedigrees from recognized kennel clubs to enter its Foundation Stock Service, it is not very helpful to landrace breeds that lack Western-style dog clubs.

One threat to these landrace breeds is the increasing presence of Western breeds in eastern and central Europe, including many protection and fighting dog breeds. Dog fighting is accepted or tolerated in many of these countries and has exploded in popularity in areas of former communist or Taliban control. Crossbreeding or encouraging aggression is totally inconsistent with the essential qualities of the measured response and the balanced, nurturing temperament found in a good livestock guard dog. It is now necessary to take great care in importing dogs from these areas where dog fighting is flourishing, since crossbreeding is being used to produce larger and fiercer fighting dogs.

GENETIC CONCERNS WITH PUREBREED DOGS

Some breeds have not been in North America long enough for breeders to document health concerns and average life spans. This does not mean that these breeds are without health concerns. Do not believe breeders who claim that they don't need to test for potential problems, such as hip dysplasia, because their breed is unspoiled, primitive, or imported. Hip dysplasia is present in all large and giant breeds, and it is unethical to ignore the opportunity to gain health information through appropriate and available screening.

However, individuals and clubs who seek to conserve rare breeds will probably find it necessary to include dogs that carry some genetic defects in order to maintain enough genetic diversity in a small population. In addition, it is difficult to predict life span in many of the landrace breeds. In general, dogs of these breeds will live ten to fourteen years, with some breeds trending to the lower or higher end of this range. If you are interested in a rare breed, take the time to speak with several breeders and educate yourself about the issues present in that breed.

Breed Registries

See the resources, page 216, for contact information for these registries.

AKC	American Kennel Club (AKC Foundation Stock Service for rare breeds)
ARBA	American Rare Breed Association
CKC	Canadian Kennel Club
FCI	Fédération Cynologique Internationale (World Canine Organization)
KC	Kennel Club (UK)
UKC	United Kennel Club

Health Registries

These registries test dogs for orthopedic or genetic diseases. Breeding dogs should have their hips tested by either the OFA or PennHIP methods.

OFA	Orthopedic Foundation for Animals
PENNHIP	University of Pennsylvania Hip Improvement Program
CERF	Canine Eye Registration Foundation

Breed Profiles

Livestock guard dogs were developed and used in a great sweep of areas from western Europe to Asia. The movement of nomadic peoples and the seasonal movements of flocks and dogs has not traditionally respected political or national borders.

Therefore, livestock guard dogs are necessarily related in function, form, and even appearance. However, people in different areas have also made choices for specific traits, resulting in standardized or landrace breeds that may differ in their behaviors and even style of work.

The breeds listed below have been standardized or recognized or are on the path to recognition, either within their homeland or in another country. Livestock guard dogs are also at work in many adjacent countries and areas. Some of these breeds are closely related types or crossbreds, but we may discover additional distinctive landrace breeds in the future. Other breeds may also achieve formal recognition in the future by the FCI or within their country borders.

The breed profiles that begin on page 119 are organized by country of recognition, moving generally from the west to the east, which helps show the relationships between breeds. The name used is the most commonly associated with the breed; alternative names are listed where appropriate. In some cases, there are internal or external conflicts about the correct breed name or the breed name may be officially changed in the future with the increased recognition of national or landrace breeds.

ALPHABETICAL LIST OF BREEDS

- Akbash Dog (p. 143)
- Anatolian Shepherd Dog (p. 149)
- Bukovina Shepherd (p. 141)
- Cão de Castro Laboreiro (p. 120)
- Cão de Gado Transmontano (p. 122)
- Carpathian Shepherd (p. 142)
- Caucasian Mountain Dog (p. 153)
- Central Asian Shepherd Dog (p. 154)
- Estrela Mountain Dog (p. 119)
- Gampr (p. 151)
- Greek Sheepdog (p. 138)
- Great Pyrenees (p. 125)
- Kangal Dog (p. 146)
- Karakachan (Bulgarian Shepherd Dog) (p. 139)
- Kars Dog (p. 148)
- Karst Shepherd (p. 135)
- Komondor (p. 132)
- Kuvasz (p. 133)
- Maremma Sheepdog (p. 127)
- KyiApso (p. 157)
- Mioritic Sheepdog (p. 143)
- Polish Tatra Sheepdog (p. 129)
- Pyrenean Mastiff (p. 124)
- Rafeiro do Alentejo (p. 121)
- Sage Koochee (p. 156)
- Sarplaninac (p. 137)
- Slovak Cuvac (p. 131)
- South Russian Ovcharka (p. 152)
- Spanish Mastiff (p. 123)
- Tibetan Mastiff (p. 158)
- Tornjak (p. 136)

GEOGRAPHICAL LOCATION OF BREEDS
(SEE ALSO NEXT PAGE)

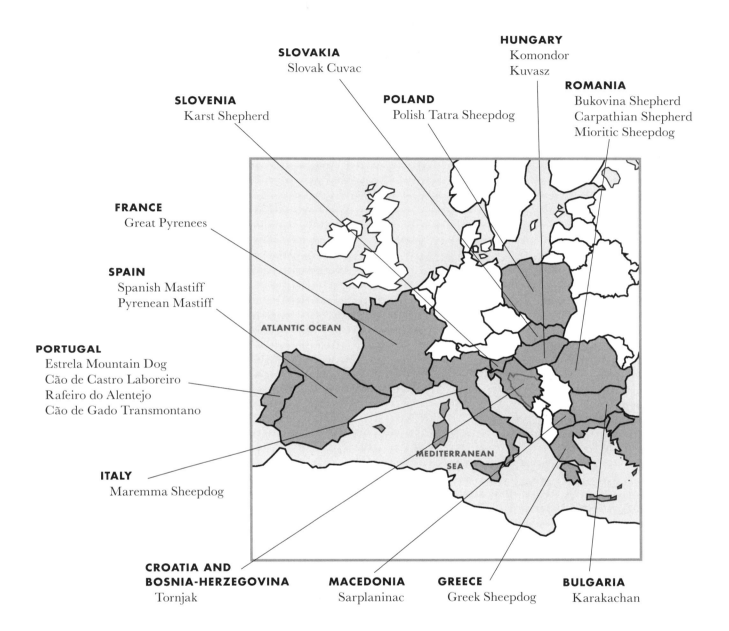

SLOVAKIA
Slovak Cuvac

HUNGARY
Komondor
Kuvasz

SLOVENIA
Karst Shepherd

POLAND
Polish Tatra Sheepdog

ROMANIA
Bukovina Shepherd
Carpathian Shepherd
Mioritic Sheepdog

FRANCE
Great Pyrenees

SPAIN
Spanish Mastiff
Pyrenean Mastiff

PORTUGAL
Estrela Mountain Dog
Cão de Castro Laboreiro
Rafeiro do Alentejo
Cão de Gado Transmontano

ITALY
Maremma Sheepdog

CROATIA AND BOSNIA-HERZEGOVINA
Tornjak

MACEDONIA
Sarplaninac

GREECE
Greek Sheepdog

BULGARIA
Karakachan

LIVESTOCK GUARD DOG BREEDS 101

GEOGRAPHICAL LOCATION OF BREEDS
(SEE ALSO PREVIOUS PAGE)

REPUBLIC OF GEORGIA
Caucasian Mountain Dog
Caucasian Ovcharka

SOUTH RUSSIA
Ovcharka

CENTRAL ASIA
Central Asian Shepherd Dog

TURKEY
Akbash Dog
Kangal Dog
Kars Dog
Anatolian Shepherd Dog

ARMENIA
Gampr

AFGHANISTAN
Sage Koochee

TIBET
KyiApso
Tibetan Mastiff

LIVESTOCK GUARDIANS

A The puppies climb out from their fleece bedding area to meet some young sheep, encountering fleece on living animals.

B This pile of snoozing animals includes a litter of pups with a young lamb joining in, as her mother lies nearby and provides supervision.

LIVESTOCK GUARDIANS 103

A When it came time for sorting the lambs from the ewes for weaning, this adolescent pup stayed with his lambs during the entire process.

B After their initial caution, these young pups became brave enough to befriend an adult horned ram.

C A young pup plays, while keeping an eye on her lambs.

A While the flock feeds, this adolescent puppy stays on alert and watches for disturbances or threats.

B This young dog shows proper attentiveness to a ewe, sitting quietly and licking her nose.

C Curling up against their bodies, a young dog beds down with his pen of lambs.

LIVESTOCK GUARDIANS

A This pack of half-grown pups demonstrates the guard dog attitude in their body language as they patrol.

B This male Pyrenees shows the battle scars of predator encounters.

A An adult dog picks a vantage point to watch over his sheep herd, as well as supervise the adolescent pups joining in guarding activities. The adult dog will discipline the pups if they get too boisterous with the sheep.

B The early relationship between the young animals is attentive while they are becoming completely comfortable with one another. This tail-up pup wants to play, but the lamb isn't having any nonsense.

C This adult male guardian stays in the middle of his flock, blending in with his charges.

Sometimes the dog will serve as the lead sheep, leading the herd out for the day's grazing or to water.

A Kangal Dog Toro demonstrates the size, power, and athletic nature of his breed.

B Wearing his iron spike collar, Cesur the Kangal Dog guards his sheep flock from wolves in Turkey. The name Cesur means "brave."

C In his homeland, the Tibetan Mastiff was bred as a guardian of property as well as livestock. Tibetan Mastiffs are well suited to cold environments.

LIVESTOCK GUARDIANS 109

A Two Anatolian Shepherd Dogs exhibit the calm and watchful behavior of mature livestock guardians.

B In Bulgaria, the good-natured Bulgarian Shepherd (called Karakachan Dog in North America) successfully combats one of the highest numbers and densities of wolves and bears in Europe.

A The most popular livestock guardian dog breed in the United States, the Great Pyrenees possesses both an essential good nature and protective instincts.

B The human-oriented Kuvasz is a very active dog, responding quickly to threats.

C In the United States and Canada, the recently imported Pyrenean Mastiff is often found working as a home and family guardian.

A The Estrela Mountain Dog is protective of home and property, as well as his flocks.

B Central Asian Shepherd Bam Bam is careful and gentle with a young Karakul lamb.

C The imposing Caucasian Mountain Dog has a distinctive bear-like appearance.

D Although less well known in North America, the Maremma is an excellent guardian of sheep or goats.

A Despite his mop-like appearance, the Komondor is one of the most serious livestock guardians.

B The giant Spanish Mastiff may appear sleepy but he serves as a powerful guardian of both home and flock.

C This shorthaired Akbash Dog has been patiently and successfully socialized to poultry.

LIVESTOCK GUARDIANS

A High in the Andes a llama surveys his flock companions.

B A guardian llama can be an enjoyable addition to the farm.

C Ears laid back in threat, this llama is obviously irritated that his herd has been approached.

This guardian llama has placed himself on a high point in his pasture where he can observe the area for disturbances.

LIVESTOCK GUARDIANS

A Llamas are highly social animals that enjoy the company of their flock mates.

B Some guardian llamas assume leadership of the flock and actively patrol fence lines.

A A ram and a yearling burro sniff noses in greeting.

B A young burro, adopted from the Bureau of Land Management, is introduced to a group of young lambs. Bonding occurs quickly.

LIVESTOCK GUARDIANS

A When the lambs wander too far from what this young guardian burro considers her safety area, the burro moves them back, with her body language saying it all. If one of the lambs gets too far from the rest, the burro will also herd the stray back to her flock.

B In areas of Switzerland where grazing pastures are open to day hikers, guardian donkeys are used instead of dogs that might be perceived as threatening.

Portugal

ESTRELA MOUNTAIN DOG
es-TRA-la
(Cão da Serra da Estrela)

The Estrela Mountain Dog is the descendant of many generations of working dogs that followed the flocks and their shepherds up into the summer grazing on the rugged, isolated mountains known as the Serra da Estrela. These sturdy guardians protected the flocks primarily from the Iberian wolf. The flocks were always accompanied by shepherds during the day and then confined at night. As the cooler fall weather set in, they returned to lowland pastures, where the flocks were sometimes left alone with the dogs during the day and were herded into fenced areas and again left with the dogs at night. The dogs sometimes became guardians of both farm and house as well.

At the beginning of the twentieth century, as foreign breeds began to come into Portugal, there were efforts to recognize this native breed and celebrate it at informal shows called *consursos*. The first breed standard was written in 1933, describing a functional dog with some special features: his tail ended in a little curly hook and his small rose ears were held folded back against his head. However, through the war years farmers and sheep raisers continued to breed their dogs to their own standards, not a written guideline. It wasn't until after the Portuguese revolution in 1974 that popular interest grew in the native breeds of the country.

Today the Estrela Mountain Dog is the most popular of the native Portuguese breeds. The Estrela is still hard at work as a flock protector and a home guardian, but it is also used as a police and military dog in Portugal and has been trained for scent work and carting as well. Increasingly the Estrela is being kept as a pet or show dog. Grupo Lobo (see page 213) has placed the shorthaired variety of the Estrela with livestock producers, achieving great success against wolf predation.

An Estrela breed club was established in the United Kingdom in 1972. Estrelas are also bred in several other European countries. Estrelas have occasionally been imported to North America, but there was not an organized effort to organize a club with recognized imports until 2001. Both the AKC Foundation Stock Service and the UKC recognize the Estrela.

APPEARANCE

The Estrela has two long-recognized and acceptable types. The longhaired Estrela is seen most frequently, but there is a large number of shorthaired dogs as well. Both are double-coated and have a harsh outer coat, capable of withstanding difficult weather. Longhaired Estrelas have a thick ruff around the neck, rather than a Mastiff-like dewlap. In both types the nose, lips, roof of the mouth, and eyelids are black. The eyes are amber in color. The ears are small and held back against the head. The low-set tail is long and ends in a distinctive small hook. When the Estrela is alert, the tail is held horizontally. Colors include yellow, light fawn through dark mahogany, brindle, and wolf gray. A black mask is highly desirable no matter the color. White markings may appear on the chest, underparts, feet, and tail. All-black, all-white, and pinto dogs are unacceptable.

Because of their substantial bone structure, Estrelas are impressive, square dogs. In height, males range from 26 to 29 inches, with females slightly shorter. Estrela males weigh 88 to 110 pounds, and females 66 to 99 pounds. Heavier mountain-type Estrelas, weighing up to 132 pounds, are also found in Portugal.

CHARACTER/TEMPERAMENT

Estrelas are protective of their family and property and prone to using their especially loud, threatening barks. As adults they become quite suspicious of strangers and sometimes are aggressive toward other dogs. However, owners report that Estrelas are particularly fond of children. The self-thinking nature of these livestock guard dogs has sometimes led to their being called stubborn. Puppies go through a very active phase but eventually settle down into the calmness typical of this breed. Estrelas are agile climbers and jumpers, so they need good fences. They must be socialized.

SPECIAL CONSIDERATIONS

Fertility problems have been noted in this breed, especially concerning the reluctance of the female to breed. Also, Estrelas were raised for centuries on marginal diets, and now it is important not to overfeed puppies. Potential health problems have not yet been documented in the United States or Canada.

CÃO DE CASTRO LABOREIRO
COW duh KAS-trew lab-var-AY-dew
(Portuguese Cattle Dog)

The Cão de Castro Laboreiro takes his name from that of a village in the extreme northern tip of Portugal. In this isolated area the Castro Laboreiro was used as a cattle guard. Due to the differences in their behavior and appearance, it was long supposed that the Castro Laboreiro was not closely related to the other Portuguese livestock guardians. In 2005 DNA studies confirmed that the breed is genetically distinct from the Estrela Mountain Dog and the Portuguese Sheepdog (Cão Serra de Aires) the native herding dog of Portugal. The Castro Laboreiro is recognized by the FCI.

The Castro Laboreiro's role has grown beyond that of cattle guard. He is also used to protect small flocks of sheep or country homes, as a police dog, and even as a companion dog. As pride has grown for the native Portuguese breeds, the Castro Laboreiro has gained attention, and in the mid-1990s breeders formed an association that registers about one hundred puppies annually. Grupo Lobo has placed Castro Laboreiros with livestock producers and documented their outstanding success preventing predation from wolves.

Dr. Ray Coppinger of the Livestock Guard Dog Project at Hampshire College imported three Castro Laboreiro dogs to the United States in 1980 for evaluation as livestock guardians. Those few producers who had to the chance to use them as livestock guard dogs were full of praise for their abilities and disappointed that the breed has not been more available for livestock protection work in North America.

APPEARANCE

The Castro Laboreiro does not look like many of the other guardian dogs, which works to his favor in acceptance by wary neighbors. Described as being similar in appearance to a Labrador Retriever, he has a short, close coat without any undercoat. He has wolflike coloring, from light to very dark. Sometimes the dog is literally colored like a wolf, lighter underneath and darker on the head, back, and shoulders. Black may have been more common in the past. Today a favorite color pattern is brindle, also known as mountain color, featuring gray, brown, or mahogany stripes in darker shades. The dog's nose, eyelids, lips, and roof of the mouth are always black.

The Castro Laboreiro is rectangular in appearance with clean lines, never wrinkled or mastiff-like in looks. Males stand 22 to 24 inches tall, with females slightly shorter at 20 to 22 inches. The Castro Laboreiro generally weighs 50 to 75 pounds, but somewhat heavier dogs are not unusual. He carries his saber tail low unless he is alert, when he carries it above the top line.

CHARACTER/TEMPERAMENT

While intelligent and a free thinker, the Castro Laboreiro is also described by owners as less stubborn and more easygoing than many other livestock guard dog breeds. He is also an active dog, given to walking and inspecting his property. In addition, the Castro Laboreiro is known for his loud, unique bark, which begins with low tones and escalates into high-pitched howling.

RAFEIRO DO ALENTEJO
ra-FAY-dew dew ah-len-TAY-joo
(Portuguese Watchdog)

The ancestors of the Portuguese Rafeiro guarded the great herds of cattle that were herded in search of fresh pastures between the southern Alentejo

lowlands and the Douro highlands in the north, a seasonal migration known as the *transhumancias* or transhumance. Wolves, bears, and humans all preyed upon the valuable stock. The Rafeiro was also used to guard farms and rural estates in the south, as well as flocks of sheep.

In the 1940s and '50s, a census was taken of the breed and an official standard was written. However, the number of dogs in the breed continued to fall, as did the quality of the dogs. In 1980, when Dr. Ray Coppinger went in search of the traditional Rafeiro for the Livestock Guard Dog program at Hampshire College, he could not locate any dogs. Since then, interest has grown in all of the native Portuguese breeds and a society was formed to preserve the Rafeiro do Alentejo. Today the breed remains rare but has a growing population and new breeders, with three hundred to four hundred puppies registered annually. Grupo Lobo has placed Rafeiros with sheep producers to combat wolf predation with good success.

APPEARANCE

The traditional belief is that the Rafeiro is a combination of the Estrela Mountain Dog and the larger, heavier Spanish Mastiff found to the east. The Rafeiro has a smooth to medium-smooth coat that can be quite dense. He has the heavy forequarters, neck, dewlap, and head of the mastiff. The Rafeiro male stands 26 to 30 inches tall, with females somewhat smaller. Males weigh 110 to 132 pounds, and females weigh 99 to 121 pounds. The tail is carried low and the ears are pendulous. The nose, lips, and eyelids are dark in color. Many colors are acceptable, including black, tan, yellow, wolf gray, and brindle. White markings can be extensive, and pinto-colored dogs are common.

CHARACTER/TEMPERAMENT

The Rafeiro is a very calm, self-confident dog, although he can be quite aggressive to threats. He has a deep and strong bark. Among his family he is known for loyalty and docility with children. The Rafeiro remains a livestock and farm or estate guardian, but not a housedog. This is a strong-willed dog that needs a firm, experienced owner.

CÃO DE GADO TRANSMONTANO
COW duh GA-dew trans-mon-TA-new

The Portuguese believe that the Cão de Gado Transmontano is closely related to the Rafeiro do Alentejo, the southern guard dog that made yearly trips up into the northern highlands of Portugal. Originally called the Rafeiro Montano or Rafeiro Transmontano, the dogs kept in the north were further bred and selected by the local peoples. As the great *transhumancias* ended in the early twentieth century, the northern and southern types became separated. The Cão de Gado Transmontano remains the guardian of the extensive sheep flocks still found in the north. More than 95 percent of the contemporary population is used for this purpose. Wolves and feral dogs are the primary predators. It is believed that no Cão de Gado Transmontano dogs have been exported outside of the country at this time.

Kennel clubs do not officially recognize the Cão de Gado Transmontano, although a breeders' association has been founded. The Portuguese government, through the Parque Natural da Montesinho, is very supportive of the use of this livestock guardian and promotes the dogs through registration of litters, the distribution of puppies to sheep owners, and an annual exhibition. Grupo Lobo also supports the placement of these livestock guard dogs for wolf protection.

APPEARANCE

This landrace population is quite homogeneous. The dogs are generally white in color with large patches of black, orange, brown, or wolf gray. The coat is medium-long with a dense undercoat. The Cão de Gado Transmontano is a large and powerful dog, standing 27 to 32 inches tall.

CHARACTER/TEMPERAMENT

The Cão de Gado Transmontano is regarded as a calm, cautious, and reserved dog. He is known as an excellent night guardian, reliable but not highly aggressive. He responds well to positive handling. Portuguese shepherds prefer to use male dogs, so males generally outnumber bitches in guarding situations. As a result, the males tend to coexist better than do some other livestock guardian dog breeds.

Spain

SPANISH MASTIFF
(Mastin Español)

Spain is the home of a great traditional transhumance that still holds a deep cultural connection for the Spanish people. In years past millions of Merino sheep and cattle moved along the *canadas*, the centuries-old, wide cattle passageways, between the summer mountain pastures in northern Spain and the warmer winter pastures closer to the Mediterranean. The livestock guard dogs also traveled with the *transhumancia*, protecting the flocks and herds against wolves, bears, and lynx.

Some experts estimate that thousands of dogs accompanied the shepherds each year. Mountain livestock guard dogs, such as those found in the Pyrenees, were combined with the traditional mastiffs, which were beloved in the Mediterranean area, to create a landrace breed specific to Spain: the Spanish Mastiff, which is the heaviest of the livestock guard dog breeds.

Over the past century only a small portion of Spain's sheep and cattle herders have continued to practice transhumance. As the large flocks of sheep diminished, so did the numbers of Spanish Mastiffs.

Some dogs found new employment as protectors of the home, others became large game hunters, and some were used by the Spanish military as guard dogs. Interest in the native breed was revived in 1970, when a club was organized, a standard was written, good breeding specimens were identified, and the breed experienced a true resurgence.

Now recognized by the FCI, the Spanish Mastiff is a national symbol and among the most popular of the dog breeds in Spain. The breed numbers about 24,000 dogs. The Spanish Mastiff has found a new role as a family dog, although he remains a powerful, challenging dog to own. Although members of the Mastin Español Club of America have imported Spanish Mastiffs from both Spain and Europe, they are still rare in North America.

APPEARANCE

The Spanish Mastiff is a giant breed, reaching heights of 28 to 35 inches. Females weigh 145 to 170 pounds. Males generally weigh 185 to 220 pounds but can weigh even more. The dog has a rectangular build with a large, broad head and very loose lips. There is ample loose skin on the head, neck, and shoulders and a double dewlap under the neck. The tail is thick and carried low unless the dog is at alert.

The Spanish Mastiff is double coated and sheds twice a year. The coat is of medium length and dense and may be of any color or color pattern. Breeders prefer whole colors, with or without a black mask, though brindles, pinto coloring, and white markings are also admired.

The Spanish Mastiff is seen in two types, the *ligero* (light) and *pesado* (heavy). The heavy type is seen most often at dog shows. The light type is the more agile and active working guardian and more capable of activity in summer heat. Some observers believe that the heavy type has lost some of the strong traits that made it a good livestock guardian and is better suited to guarding estates or homes.

CHARACTER/TEMPERAMENT

The Spanish Mastiff has his own style of guarding. He appears more passive and sleepy than the other livestock guard dog breeds, making his presence known by his sheer size and his deep bark or growl. He does not react quickly, but if he feels his territory is being threatened, he will attack the intruder with ferocity. He is deeply suspicious of strange animals and people. He is an aloof dog, not overly affectionate, but trainable by a firm owner. This breed requires a serious, committed owner.

SPECIAL CONSIDERATIONS

The Spanish Mastiff grows rapidly as a puppy but is slow to mature. A male is not fully mature until two and a half to three years of age. Health problems in the breed include hip dysplasia, heart problems, entropion, bloat, and panosteitis in puppies. Heat and high humidity can be a problem for the massive dogs. Females often require Cesarean sections to give birth. Finally, as with most heavy mastiffs, the Spanish Mastiff is a champion drooler and snorer.

PYRENEAN MASTIFF
PEER-uh-nee-un

(El Mastin del Pirineo)

The Pyrenees Mountains have long formed the natural boundary between the Iberian Peninsula and France. Although older than the Alps, the Pyrenees are noted for the rarity and high elevations of the natural passes through them. Just south of the Pyrenees, from Aragon to Navarre, is a region rich in history and tradition. Sheep once grazed the southern slopes of the mountains here from summer through early fall, with Pyrenean Mastiffs guarding the flocks from wolves and bears. These dogs were also the favored guardians of both castle and home in the old kingdom of Aragon.

The beloved *mastino*, as the mastiff is affectionately called, faced a serious crisis in the twentieth century. First the large predators were exterminated from the Pyrenees, which lessened the need for large guard dogs. Then the country suffered through the Spanish civil war and its terrible toll on the economy of the country. The people simply could not afford to feed the big dogs, and their numbers went into serious decline.

In the 1970s a group of *mastino* admirers searched the countryside for Pyrenean Mastiffs and were able to identify about one hundred excellent specimens. A club was formed and the dogs were registered and then carefully bred and selected for the traditional type. Today there are about four thousand Pyrenean Mastiffs in Spain, and the breed is recognized by the FCI. Though these mastiffs are used primarily as a property guardians and family companions, there is renewed interest in using them as working dogs since the return of wolves to the Pyrenees. The breed has been exported to various European countries and the United States. Since 1996 members of the Pyrenean Mastiff Club of America have bred about 350 dogs in the United States and Canada, using the standard and health requirements of the parent club in Spain.

APPEARANCE

The Pyrenean Mastiff clearly shows the influences of its two related breeds, the Great Pyrenees to the north and the Spanish Mastiff to the south. Less refined and larger than the Great Pyrenees, the Pyrenean Mastiff is broader in the body with a wider head and large dewlaps. The face bears close resemblance to that of the Spanish Mastiff, with its pendulous lips and sleepy look. The ears tend to fold back into the coat. The coat of the Pyrenean is less profuse than that of the Great Pyrenees but longer than that of the Spanish Mastiff.

The Pyrenean Mastiff is a very large dog. The average height for males is 32 inches, with females reaching 29 to 30 inches. There is no maximum height limit for the breed, and a tall, well-proportioned dog is much appreciated. The Pyrenean weighs 120 to 150 pounds or more. It is a rectangular dog with an impressive head and loose skin around the neck. The dewlap is well developed, giving an impression of great size and power. The tail hangs low and is slightly curved but never carried over the back.

The thick coat is medium-long and coarse, not soft or woolly. The base color is always white, with well-defined patches in several colors. A well-defined mask is appreciated, but puppies are often born with mismarked masks. The dog is double-coated and requires some grooming to prevent matting.

CHARACTER/TEMPERAMENT

The Pyrenean Mastiff is a more active dog than the Spanish Mastiff, but he is also calmer, less sharp, and more suited to family living, although he is capable of strong reaction to threats when aroused. Although his bark is deep, the Pyrenean does not seem to bark as much as other livestock guard dogs, which may endear him to neighbors. He is fond of children and lacks the hair trigger of many other guardians. Although suspicious of strangers, the Pyrenean will accept visitors who are introduced properly. In the United States and Canada, many of the newly imported Pyreneans are working as home and family guardians.

The Pyrenean Mastiff has a hearty appetite and eats more than many other livestock guard dogs. He is also an excellent fence climber and needs to be safely confined. Other than typical puppy chewing and digging, the breed is not known for these irritating habits in adulthood.

SPECIAL CONSIDERATIONS

Health concerns include both entropion and ectropion and a tendency toward conjunctivitis. The expression of hip dysplasia has been controlled due to the strict use of screening X-rays conducted through the European Radiology Association for both European and American dogs. Puppies grow very rapidly and can be prone to panosteitis. As with many giant dogs, bloat can also be a concern, and Pyreneans may drool in hot weather or when they are excited. Humid climates are uncomfortable for Pyrenean Mastiffs, although they can handle dry heat by retreating to cool dens.

France

GREAT PYRENEES
PEER-uh-nees
(Pyrenean Mountain Dog, Chien de Montagne des Pyrenées)

The Great Pyrenees is the most familiar livestock guard dog breed to the general public, even though most Pyrs do not work as livestock guardians in North America, Europe, or even their homeland in the Basque country of the Pyrenees. There is no doubt about this breed's ancient role as a fierce protector against wolves and bears. Its popularity

Great Pyrenees

spread as royalty took Pyrs into their estates and homes as protectors. Louis XIV declared the Great Pyrenees the national dog of France. General Lafayette sent a pair to a friend in America. Queen Victoria owned one. The breed's popularity extends far beyond France, and the Pyrenean Mountain Dog, as he is called outside North America, is now known around the world.

Although the breed is probably related to the Pyrenean Mastiff, it has been selected for its particular appearance and behavior for quite a long time. Fanciers began to write standards for their beloved breed at the beginning of the twentieth century, just at the same time that he was losing his traditional job in the mountains because wolves had been exterminated. World War I decimated Pyrenees numbers in their homeland. The Reunion des Amateurs de Chiens Pyreneans, the club that still champions all native breeds of the Pyrenees, came to the rescue then and again after World War II, when, as was the case for many of Europe's large breeds, the population of the Pyrenees fell to dangerously low levels.

Breeding dogs were exported to North America before World War II and recognized by the AKC in 1933. The Pyrenees continues to have considerable popularity today, with more than two thousand puppies registered annually.

APPEARANCE

A large factor in the breed's popularity is probably due to its beauty. The French describe the ideal head as resembling that of the Pyrenees brown bear, with very little stop between the muzzle and the head. The Pyr has kindly, almond-shaped eyes sometimes described as dreamy. The ears are small to medium in size, low, and held close to the head. The effect is indeed bearlike. The Pyr should not have the heavy flews, blunt muzzle, or dewlap of a mastiff. The plumed tail is carried low unless the dog is on alert, when it can form a wheel over the back. The Pyr also has double dewclaws, which must not be removed according to the standard.

The Pyrenees is a large dog, with a height of 27 to 32 inches for males and 25 to 29 inches for females. Females weigh 85 to 110 pounds and males 100 to 140 pounds. The Pyr appears larger than he is, due to his immense coat. The long outer coat is weather resistant and sheds dried mud. The fine undercoat is shed at least once a year. The coat forms a ruff around the neck, especially in males. Although pure white dogs are frequently used to represent the breed, white dogs with patches of light tan, gray, or red are more common. Badger patches on the head are especially attractive. Many of these patches appear in puppies but fade as the dogs mature.

CHARACTER/TEMPERAMENT

The Pyr is generally a calm dog, slightly aloof around strangers and independent in nature. A strong breed trait is the Pyr's love of children and nurturing behavior toward young animals. Socialization and training are requirements, despite the Pyr's essential good nature. You cannot expect your Pyr to calmly welcome an unwanted visitor; he is still a guard dog at heart. Pyrenees are often described as nocturnal, and they do bark at disturbances during the night. They can also be fence climbers and have a strong urge to roam.

When efforts began in North America to utilize the traditional livestock guarding breeds, the Pyrenees was the only breed already well established on the continent. The Pyr did well in early tests and proved itself the least aggressive to humans and stock. In the Great Pyrenees, unprovoked human aggression is considered a serious fault and good temperament is a virtue.

Many Pyrenees work reliably today as either farm and home protectors or traditional livestock guard dogs. The Pyrenees works through barking and warning threats, making physical contact only if an intruder persists. He generally keeps himself at a distance from the flock to keep watch. He is often quite lethargic during the day and more active at night. Pyrs often dig dens, especially in hot weather, to find relief from the heat.

SPECIAL CONSIDERATIONS

Since so many Pyrenees are raised primarily as companion dogs, it is advisable to search for breeders who specialize in livestock guards or who pay careful attention to good guardian qualities in their breeding choices. If the parents are not working dogs, ask whether any offspring have been successfully placed in working homes. Some lines have been bred primarily for show or companionship and may lack proper working temperament.

Great Pyrenees breeders advise that the coat not be clipped or sheared, since the dog needs it for protection from the sun. In reality, the old working dogs of the Pyrenees Mountains often appeared in dirty, tangled, and matted coats. Unless you are prepared to groom your dog weekly, especially during the spring molt of the profuse undercoat, the coat will mat and foreign matter, especially burrs and foxtails, will become stuck in the coat. Some owners do give the dog a trim each spring.

Dogs are usually available through the rescue efforts of the national and regional Pyrenees clubs. Some are potential working livestock guardians, and these groups can be contacted for advice and recommendations. Often these dogs were surrendered to rescue because they did not do well in an urban environment. Nighttime barking is the primary reason for surrender. Unfortunately, there are also large numbers of crossbred Pyrenees dogs in shelter situations. The most common crosses are Anatolian/Pyr and Akbash/Pyr, since these three breeds are the most common dogs in ranch situations today. Most of these crossbred dogs are surrendered due to a variety of unsuitable behaviors.

Health concerns include hip dysplasia and other orthopedic problems, bloat, thyroid disorders, deafness, progressive retinal atrophy, and skin problems. The good news is that many of these conditions can be detected by testing and good breeders can avoid them. The Great Pyrenees Club of America requires that breeding dogs be tested for hip dysplasia and that these scores be made available to puppy buyers.

Italy

MAREMMA SHEEPDOG
mare-emma
(Il Pastore Maremmano Abruzzese)

The Maremmano Abruzzese sheepdogs were an integral part of the ancient pattern of transhumance in Italy, and ancient Roman writers often mentioned these white dogs. In Italy the Maremma encompasses the plains and coastal western land down to Rome, where the climate is mild in the winter and there is good grazing. As the grass withered in the summer heat, the sheep and goats were moved up into the cooler Apennine Mountains of the Abruzzi. In the mountains the Maremmano Abruzzese,

Maremma Sheepdog

working in groups of three or more, faced predators such as wolves, bears, and feral dogs. In the evenings the sheep were usually penned in enclosures along with the dogs. The dogs accepted the daily milking routine of the sheep and goats and tolerated the ten-day trip through villages on the traditional roads or *tratturi*. In many ways these dogs were more bonded to their stock than to their territory.

As sheep herding declined, more Maremmano Abruzzese began to be kept as majestic estate guardians, especially in Tuscany. Queen Victoria owned a pair, and the British Kennel Club recognized the breed in 1936. In the Maremma today the sheep are moved by truck, but the dogs are still necessary to protect against predators. With the reappearance of the wolf in the Alps, there is renewed interest in other countries in the Maremmano's guarding abilities. Originally there were various types of regional Maremmanos, and to a certain extent those types still exist today. In the 1950s, however, the decision was made to combine the Abruzzi and Maremma types into one recognized breed. The breed population is now about seven thousand dogs.

In the 1970s Dr. Ray Coppinger of Hampshire College imported some Maremmas, as they are called in the United States, for the first extensive studies on the use and effectiveness of livestock guard dogs. Maremmas proved to be one of the most successful guardians in these studies. The Maremma Sheepdog Club of America was created to register these dogs and provide support for their owners. To date the club has registered more than twenty-five hundred dogs.

Maremmas are one of the secrets of the livestock guard dog world, except to their hundreds of owners who need a highly protective, fierce guardian against serious predator threats.

APPEARANCE

The Maremma is a white dog with some shading of cream, ivory, pale orange, or lemon. The skin may be pink with dark spots. The long coat is somewhat harsh and sheds dried mud. In winter the Maremma has a thick undercoat that is well suited to extreme weather. The summer coat is much thinner and can have some color changes. The Maremma sheds heavily twice a year. The lips, nose, and eye rims are black and the eyes are dark. The Maremma has dewclaws on all four feet, which are not removed. The tail is carried low when the dog is relaxed but rises when he is alert, although it is not carried in a curl over the back. The Maremma stands 25 to 30 inches tall and weighs 70 to 100 pounds, although some individuals are heavier.

CHARACTER/TEMPERAMENT

The Maremma is an unusually calm dog that springs suddenly to ferocious action if a threat materializes. Experienced breeders sometimes call this "The Charge." Two or three Maremmas are capable of turning back bears or wolves at a pasture perimeter. Both sexes are good guardians. Maremmas patrol their perimeters and leave scent marks to discourage canine predators, but they spend most of their time with their stock or observing them from a higher spot. The Maremma is sometimes described as introverted and he often appears aloof. Owners report that their dogs are affectionate and loyal to them but happiest when they are out with their stock. The Maremma is often described as "parental" with its stock.

Due to his heritage the Maremma tends to be closely bonded to his flock, and this lessens the tendency to wander. He is especially well suited to farm flock situations where he will interact with the farmer through the daily routines. A young puppy will bond strongly to his people, stock, and home, although it is possible to bring a mature guardian onto your property with guidance from a good breeder who raises working dogs. A puppy from four to sixteen weeks of age should be raised with stock and given human attention only at feeding times. After he is fully bonded to stock, you are free to give him much more attention. Care needs to be taken until about eighteen months of age that the young dog does not develop bad habits of wandering or playing with stock. A young dog may chew, dig, or bark excessively. Maremmas are mature guardians by age two.

The Maremma Sheepdog Club of America has written an excellent guide specific to raising and training a Maremma.

SPECIAL CONSIDERATIONS

The Maremma Sheepdog Club of America seriously discourages the keeping of this breed as a pet. Their caution is strongly worded: "The pet Maremma, without a flock and large area to guard, will gradually become more possessive of its bonded humans and of its limited territory, and more defensive of its possessions, while it will also become less discriminatory between friend and foe." Ninety percent of the dogs that come to the club's rescue effort are from pet homes. The club goes on to warn that even the most socialized dogs will need to be introduced to all strangers and may still not accept them as visitors in the future, unless you are present.

Some lines of North American Maremmas have been selected to be very aggressive. Others are more stable and reliable. There is some hip dysplasia in the breed, as well as sensitivity to anesthesia. The heavy coat retains flea and tick sprays, so be cautious about overuse of these products.

Poland

POLISH TATRA SHEEPDOG
TAT-rah
(Polski Owczarek Podhalanski)

Poland lies across the beautiful Tatry Mountains from Slovakia, and the lovely Polish Tatra is obviously related to the Slovak Cuvac. For many centuries shepherds all across the Carpathian Mountains took sheep and their Tatra guard dogs up into the summer pastures. While two or more dogs watched the sheep, the shepherd's days were

busy with milking and making cheese. The Tatra is noted for his style of protection against predators: He will place himself between the flock and predator, remaining close to the flock while barking to warn the predator and alarm the shepherd. The Tatra attacks only if the predator persists and moves in close to the flock.

In the 1930s a standard was written for the breed, and it began to receive some international recognition. World War II seriously endangered the breed. In 1954 the Communist government confiscated all private property in the highlands and created a national park in the Tatry Mountains. The shepherds were displaced to other sites in the Carpathian Mountains and sheep keeping in the area declined. Ironically, the Tatry mountain meadows are now overgrown with weeds, and park authorities are hoping to reestablish summer grazing. Shepherds in other areas are still using Tatras as livestock guard dogs with documented success, and their use is now encouraged as large predators return to the Carpathians.

In Poland, when the sheep returned to lowlands in the winter, the Tatra was used as a home and farm guardian, as well as for draft work hauling small carts for the dairies. The Tatra also worked as a drover, helping to move the flocks along, and sometimes accompanied men on patrol in the mountains.

Fortunately, lovers of the Tatras rallied around them in the 1960s, and the breed received FCI recognition. There are now approximately two thousand Tatras worldwide, primarily in Poland, but also in Germany and Britain. They were imported to North America beginning in the 1980s, and the continent currently has a population of about three hundred dogs.

APPEARANCE

The Tatra male stands 26 to 28 inches, with the female slightly shorter. He weighs 80 to 130 pounds. His ears and tails are not cropped. He is a heavy dog with a massive appearance, partially due to its abundant and heavy coat. The coat is always pure white, accented by a black nose, black lips, and black eye rims. The outer coat is quite weather resistant and sheds dried mud. The undercoat is profuse and was actually once spun as yarn in Poland. The coat does require grooming. Outdoor dogs shed twice a year, but dogs kept indoors will shed year-round.

CHARACTER/TEMPERAMENT

Perhaps due to his history of use, the Tatra is an even-tempered dog that shows great restraint and lacks the short trigger of some other livestock guard dog breeds. The Tatra is well suited to small farms that need a livestock guardian that will accept visitors. Because he is a social and affectionate dog, he is not well suited to large range situations. Tatra owners in North America report that their dogs are willing herding assistants as well as guardians. They are especially successful with sheep breeds that flock tightly. Tatras will also work well with other dogs, including herders. They do not drool. Their life span is ten to twelve years.

SPECIAL CONSIDERATIONS

The Polish Tatra Club of America requires DNA typing on all of its dogs, as well as mandatory OFA hip screening for dogs used in breeding. The club has reported a few cases of bloat, patellar luxation, and allergic dermatitis. Although the Tatra makes a good family companion, he must be socialized, requires exercise and good fencing, and will bark at disturbances, especially at night.

Slovakia

SLOVAK CUVAC
CHEW-votch
(Slovensky Cuvac)

The Slovak Cuvac is closely related to the Polish Tatra to the north and the Hungarian Kuvasz to the south. The Carpathian Mountains stretch across most of its native country, with the Tatra peaks reaching the highest. In times past the Slovak Cuvac accompanied sheep flocks taken up into summer pasture camps, or *salase*, by shepherds who milked and made cheese daily. The dogs defended the flocks against wolves and bears, though these predators were hunted into very low numbers during the first half of the twentieth century. Traditionally the dogs were also used to assist with herding and droving.

By the 1920s the numbers of the Slovak Cuvac had fallen, although a standard was written for the breed in 1920 and a club formed in 1930. World War II intervened, industrialization progressed, and the practice of summer grazing changed. By the 1950s and '60s, Cuvac livestock guard dogs were generally chained at the farm or summer pasture, primarily to warn their owners against human thieves, and livestock owners generally hired workers to work at the *salase*. This is still the situation today.

However, Slovak Cuvac dogs found new homes as family and house protectors. FCI recognition of the breed came in the late 1960s, and the breed has since been exported to other European countries and North America. There are about two thousand registered Slovak Cuvac dogs in Slovakia, and slightly fewer than forty dogs have been exported to the United States.

With the resurgence of large predators in the Carpathian Mountains, there has been a renewed interest in the traditional use of livestock guard dogs. Unfortunately, the knowledge of how to socialize dogs to the sheep and patiently train them through their adolescence has been lost. There are now ongoing efforts to reestablish the Slovak Cuvac as a true working livestock guardian by the Slovak Wildlife Society and Born Free. With proper training and working conditions, the breed has shown it still possesses good instinctive traits.

APPEARANCE

The Slovak Cuvac is less massive than the Polish Tatra and slightly shorter than the Kuvasz. Males average 24 to 28 inches in height, and females slightly less. Weights range between 80 and 100 pounds. The Slovak Cuvac has an abundant, medium to long, wavy coat. The undercoat is particularly dense and sheds heavily. The profusely haired tail is carried low unless the dog is at alert. Males have a distinct ruff of hair around the neck. This breed requires regular grooming to prevent mats. The color is white, with yellow shading allowed on the ears. The nose, mouth, eye rims, and palate are black.

CHARACTER/TEMPERAMENT

The Slovak Cuvac is noted for his affectionate nature toward his family and especially children. He remains a livestock guard dog at heart, however, and requires socialization and a dedicated owner. He is also a very agile and fast dog that requires daily exercise.

Hungary

KOMONDOR
KOH-mohn-dor

The Puszta, a vast grassland plain that extends across most of Hungary, is the home of two old livestock guard dog breeds: the Komondor and the Kuvasz. The dreadlocked Komondor is thought to have come to this plain with the Cuman people in the thirteenth century. *Komondor* means "dog of the Cumans"; its plural is *Komondorok*. Traditionally the dog lived out on the grasslands from April to November. The long, heavy cords and felted plates of his coat were reputed to protect him from the weather and the wolves. He was rarely taken into the cities as a house guard, being kept instead in the rural areas. Despite his comic, moplike appearance, the Komondor is one of the most serious and imposing livestock guard dogs. Hungarians know better than to approach a Komondor with his sheep or in front of his house.

Komondorok began to find their way to Europe in the early 1900s. The AKC recognized the breed in 1937, but during the war years and the early years of Communist rule, fanciers had difficulty establishing the breed in North America. The Komondor also suffered in its homeland, where it remains rare today. Imports to North America were resumed from Hungary in 1962. Today there are a thousand to fifteen hundred Komondorok registered in the United States, but only seventy to eighty puppies are registered each year. The Komondor is not widely used as a livestock guardian in the North America, partially due to its rarity but also because of its formidable nature. The Komondor is now also recognized by the UKC.

APPEARANCE

Komondor males stand at least 27 inches tall, with females at least 25 inches. Males are typically 80 to 100 pounds, with females slightly smaller. Under all of that hair, the Komondor is medium boned but strongly muscled. The skin is gray, but the nose and lips are black in color.

The coat is naturally white, although even with extreme care it is nearly impossible to keep the cords of a mature dog pure white since he does not shed his coat. Cream or buff shading in puppies usually fades out. The soft coat of puppies begins to form mats at six to eight months of age. The soft undercoat actually becomes trapped in the coarse outer hairs. The owner usually has to help the cords form by splitting the mats into quarter-size cords. This must be done frequently until the cords have developed; otherwise large mats, called plates, will develop. Do not expect the coat of a working Komondor to look like that of Komondorok prepared for showing.

The Komondor's coat is best suited to drier climates. Moisture can lead to hot spots, odor, and mildew. Dogs will scratch out cords if fleas or ticks bother them. Dogs must also be checked regularly for long hair between the pads of the feet and in the ear canals. Working Komondorok are often clipped yearly, much like their sheep. Some owners keep the

cords clipped to about 4 inches in length and trim the hair around the mouth. The coat may also be combed out free of cords, but this is a tremendous grooming task.

CHARACTER/TEMPERAMENT

The Komondor works by staying close to his flock and observing the area; however, when an intruder appears and refuses to leave, the dog will throw himself into a fury to knock down or chase away the threat. The Komondor is extremely territorial and will guard everything he regards as his. He will be very protective of his children but not tolerant of strange or teasing children. He will bark at every disturbance. The Komondor is an active and exuberant puppy that matures very slowly. As an adult, he is not prone to wander or chase predators but is an excellent fence climber. All livestock guard dogs must be socialized, but in the case of the Komondor, if he is not heavily socialized it will be nearly impossible for strangers, such as a veterinarian, to handle him.

Komondorok like an orderly routine. Boundaries and inappropriate behavior, such as mouthiness, must be dealt with early and consistently. Owning this dog is a serious responsibility. Experienced owners caution that this breed does not respond well to harsh discipline.

SPECIAL CONSIDERATIONS

Health considerations include hip dysplasia, bloat, entropion, and juvenile cataracts. The Komondor Club of America requires all breeding stock to be certified free of hip dysplasia and genetic eye problems. Remember to make allowances for the weight of the full coat when using anesthesia or medications prescribed by weight. Flea preparations can be held in the coat, so be cautious of overuse.

KUVASZ
KOO-voss

Hungarian experts believe that the name *Kuvasz* (the plural is *Kuvaszok*) comes from that of the Chuvash, one of the ancient indigenous peoples of Eurasia. Many of the Hungarian words related to agriculture have a Chuvash origin. Although the original job of the Kuvaszok was as flock guardians, they were also taken into villages and homes. The most famous devotee of the Kuvasz, King Matthias (1458–1490), raised hundreds of Kuvaszok in his kennels, both as personal protectors and as hunting dogs. The Kuvasz remained a popular dog of royalty in Hungary and beyond, admired for his beauty and noble appearance. The breed is depicted in many works of art and literature from Hungary, where it remains the most popular native breed.

Organized breeding of the Kuvasz began in the late nineteenth century, with the first standard written soon thereafter. The FCI accepted the Kuvasz in 1934; it was the first livestock guard dog from middle Europe the organization recognized. By this time the Kuvasz was not primarily occupied as a

flock protector but also worked as a farm or estate guardian. World War II and the unsettling years of Communist control severely affected the breed. Following the war, estimates of the surviving population of purebred Kuvaszok were as low as twelve dogs. Efforts to save the breed took place in both Europe and Hungary.

Both the Kuvasz and Komondor were bred in Germany as early as 1922. Through a misconception that the Kuvasz was not to have a traditional wavy coat, and during the years of separation and isolation from Hungary, a different type developed in Germany that looked more like a Great Pyrenees. Nonetheless, it remained a fashionable breed in western Europe and was exported to North America. The AKC initially accepted the Kuvasz in 1931, although it was difficult to maintain the breed until more dogs, primarily from Germany, could be imported in the 1960s. In 1974 a new standard was written and the breed was readmitted. With the fall of Communism, the ability to exchange stock and information improved. Efforts are now under way by some breeders to correct mistaken trends in Kuvasz selection, and there is renewed interest in selecting for the traditional Hungarian standard. Both types and a range between them can sometimes be seen in the show ring. The AKC registers fewer than two hundred Kuvaszok each year.

APPEARANCE

The Kuvasz is a medium-boned dog with a lithe or lean appearance, compared to many other livestock guard dogs. One of his most notable features is his lovely, wedge-shaped head with almond-shaped eyes and black eyelids, nose, and lips. The Kuvasz should not have loose lips, a blunt muzzle, or large, broad head. Another Kuvasz feature is his wavy or undulant coat, which should not form cords. The medium-length white coat can be cream or ivory in color. The outer coat is coarse, but the breed also has an undercoat that it sheds twice a year. The plumed tail is carried low, but it will rise level when the dog is alert. The average male is 28 or 29 inches and about 90 to 115 pounds in weight. Females are 26 to 28 inches tall and 70 to 90 pounds.

CHARACTER/TEMPERAMENT

The Kuvasz is a very active, fast, agile, and athletic dog. As a livestock guardian the Kuvasz generally works at a distance from the flock, observing potential danger. He is quick to respond to threats. The Kuvasz is often described as a one-family dog, affectionate and loyal to those he regards as his. Some Kuvasz breeders will not sell a Kuvasz to a home where he will be a full-time working guardian, believing he will not be happy unless he has the opportunity to develop a strong human bond. Because he is so human oriented, the Kuvasz makes an excellent flock and farm protector, although he remains an active, aggressive dog who is suspicious of strangers. Experienced owners will remind you that the Kuvasz is a protector but not a playmate for your children. Whether kept in the field or in the home, the Kuvasz will require socialization and control.

SPECIAL CONSIDERATIONS

If you are choosing a Kuvasz as a livestock guardian, seek out a breeder who exposes puppies to farm stock and is able to choose the most appropriate dog for your situation. Some pups have a stronger chasing drive or need for human company, while others exhibit the appropriate behaviors around livestock. If the puppy is to be a full-time guardian, breeders recommend that he be kept with stock until four months of age, with minimal human contact, so that he can bond with his animals rather than people.

Slovenia

KARST SHEPHERD
(Kraski Ovcar, Krasky Ovcar, Krasevec, Istrian Sheepdog)

The Kras is a plateau that extends from southwestern Slovenia into Italy. The unusual landforms found here, composed of exposed limestone marked by sinkholes and underground rivers and caves, were given the German name *karst*. This term is now used worldwide to describe this type of geology. In 1689 historian Janez Vajkard Valvasor published a profusely illustrated multivolume work of natural history, *The Glory of the Duchy of Carniola*, in which he wrote a detailed description of the native livestock guardian dogs in this area. Today, Karst Shepherds, the descendants of these dogs, are considered a national treasure in Slovenia and an object of great pride.

In 1939 the native breed was classified together with the Sarplaninac as the Illyrian Sheepdog. Slovenian experts continued to promote the historical, physical, and behavioral differences of the two breeds until they were recognized separately by the FCI in 1968. The Karst Shepherd breed enjoys great support from breeders who worked hard to restore it following the ravages of wars and economic problems. Since Slovenian independence in 1990, the Karst Shepherd has become a symbol for the country. Today there are more than one thousand Karst Shepherds in the country as well as exports to Europe. There is also an active Slovenian Kennel Club.

APPEARANCE

The Karst Shepherd is closely related to the Sarplaninac and the Balkan sheep guardian dogs. Although some members of the breed still work with sheep, most Karst Shepherds are family companions. Physically, the Karst is a smaller and lighter dog than the Sarplaninac. Karst Shepherd males range from 22 to 25 inches, with females slightly smaller. Weight ranges from 58 to 88 pounds. The Karst Shepherd has a friendly, pleasing countenance and a compact appearance. The lips are tight and show no mastiff influence. The Karst has a double coat with a medium-length, thick, harsh outer coat that forms a ruff around the neck. The iron gray color may show shades of silver gray to very dark gray. Sometimes there is striping on the legs as well as a dark mask.

CHARACTER/TEMPERAMENT

The Karst Shepherd is a stable dog that responds well to consistent training. He is recognized for his affectionate personality and is described as being quite playful with his family. Less stubborn than other livestock guard dogs, the Karst makes a good family protector and watchdog so long as he has access to a large yard or room to exercise. With its abundant coat, the breed enjoys colder weather but would probably suffer in hot, humid climates.

Croatia and Bosnia Herzegovina

TORNJAK
TORN-yack
(Hrvatski Ovcar)

Several religious documents and other papers from the ninth through the fourteenth centuries describe the large, bicolored mountain dogs of Croatia. These dogs are no doubt the ancestors of the dogs native to Croatia and Bosnia-Herzegovina and the greater Balkans. In recent times these dogs were named Tornjak Croatian for "guardian of the pen."

Modernization after World War II led many people in what is now Croatia and Bosnia-Herzegovina to leave rural agricultural lives for the cities. Many of the sheep flocks and the dogs that watched them were eliminated. As early as the late 1960s there was concern that the traditional livestock guardian dogs were being lost, and efforts were made to locate and stabilize the population. The dissolution of Yugoslavia, followed by the terrible strife and turmoil in the area in the early 1990s, resulted in the near extinction of this landrace breed. With the creation of independent Croatia and then Bosnia-Herzegovina, there was renewed interest in the native dogs. With the resurgence of wolves and bears and a return to traditional sheep and cattle keeping, there was also a need for livestock guardian dogs.

Lovers of this breed in Croatia and Bosnia have worked hard to restore it, and it is achieving popularity in Europe as well. Over two thousand Tornjaks have now been registered. A few Tornjaks have been imported into North America.

APPEARANCE
These shepherds' dogs have always been described as white with bi- or tricolored patches. It is traditionally seen in two types, one in Croatia and another in Bosnia. In Croatia the Tornjak is a lighter dog without heavy mastiff features. In Bosnia the Tornjak is a heavier-boned, larger dog that clearly shows mastiff influence. Neither is particularly heavy or coarse. The lips are not pendulous and there are no dewlaps. Both types have a long, dense white coat with patches of solid color in black, brown, gray, fawn, yellow, or red. The dogs with lots of color and distinct head markings are the most favored. (In the past shepherds would selectively breed Tornjaks for certain colors to help them identify their own flocks from a distance.) The Tornjak is double coated, and the hair forms a ruff around the neck. He carries his plumed tail high over his back when alert.

In 2005 the two types were recognized collectively by the FCI as one breed. The standard describes a square breed with males standing 26 to 27½ inches tall and females 24 to 26 inches tall. Tornjaks generally weigh 77 to 110 pounds.

CHARACTER/TEMPERAMENT
For a livestock guard dog, the Tornjak is exceptionally trainable and social, although he does require a firm owner and socialization. He is also calm and generally friendly with people. The Croatian type

is not aggressive toward other dogs, while the Bosnian type is somewhat more dog aggressive, though he still allows shepherds to work herding dogs with their flocks and socializes well with other family-owned dogs. The Tornjak is gaining popularity as a family and home guardian, but 90 percent of the registered Tornjaks are still working dogs. The Tornjak often works in groups of two or three dogs. Puppies are socialized with the flock from birth until five months. At that time they are separated from the flock to grow up, and they are not returned to the sheep until they are adults.

Macedonia

SARPLANINAC
shar-pla-NEE-natz
(Illyrian Sheepdog, Ovcarski Pas)

The Sar Planina Mountains lie in the northwest of Macedonia, with Serbia further to the north. In the spring the sheep flocks make their way up from the Polog Valley to the summer grazing in these mountains. Where you find sheep and wolves, you find a good livestock guard dog. In this case, it is the Sarplaninac, a rough-coated mountain breed.

The Sarplaninac is a member of the group of very similar livestock guard dogs found throughout the Balkans. The close relations among these dogs, including the Greek Skilos tou Alexandros, came about in part because of the extensive traditional transhumance in the area. In 1939 the Sarplaninac and the Karst Shepherd were jointly recognized by the FCI as the Illyrian Sheepdog, Illyria being the ancient Roman name for this area. In 1957 the name of the breed was changed to the Yugoslavian Mountain Dog; later, in 1968, the Karst Shepherd was recognized as a separate breed. Following the breakup of Yugoslavia, the breed was renamed the Sarplaninac.

In the 1970s it became legal to export Sarplaninac dogs, and they soon found their way to Europe, Canada, and the United States. Sarplaninacs were successful in trials of livestock guard dogs in the United States, but they did not achieve the popularity of some of the other tested breeds. Today there are several independent breeders of Sarplaninacs in the United States, but no established clubs. The Sarplaninac is recognized by the UKC.

In its native countries the Sarplaninac has endured political and economic trials as well as much warfare. Today there is resurgence of interest in the breed, although there is some controversy between supporters of one type or another. There is also some danger to the breed from the introduction of other bloodlines; Caucasian Ovcharkas have been interbred with Sarplaninac, for example, and the Yugoslavian Army bred a military type of Sarplaninac after World War II, using some German Shepherds and Caucasian Shepherds in their breeding programs.

APPEARANCE
Slightly smaller than many other livestock guard dogs, the Sarplaninac is still a heavy-boned dog with an abundant coat and large teeth. The dog appears

larger than he actually is, with males standing 24 to 26 inches and females at 22 inches and up. Males weigh 77 to 99 pounds, and females are slightly smaller. Larger dogs are also sometimes seen, more often in the city or show ring.

The double-coated Sarplaninacs are solid in color from white to very dark brown, with iron gray a favorite. Small white marks are allowable, but not pinto or bicolored coats. Coats on working dogs are often shorter than the coats of dogs seen in the show ring. The skin must be dark, never pink. The plumed tail is slightly curved when relaxed but may rise above the back when the dog is alert.

CHARACTER/TEMPERAMENT

The real working Sarplaninac is a good livestock guardian. He is a strong-willed, intelligent, highly protective dog that shares many of the common livestock guard dog traits: wariness of strangers, aggression toward other dogs, and a somewhat aloof nature. His exercise needs are high, and he will be most happy in the country with a job to do. He is not a dog for the city. Although he will be loyal to his family, especially children, he must be heavily socialized and kept well fenced. Due to his heavy coat, he is not suited to extremely hot and humid climates.

Greece

GREEK SHEEPDOG
(Hellenikos Poimenikos)

The sheepdogs of Greece have long been famous for their fierce protection of their flocks in the mountains and their attacks upon travelers — now hikers, backpackers, and mountain bikers. In ancient times vast flocks of sheep once moved in seasonal transhumance throughout the area. Greek Sheepdogs protected the flocks from wolves and bears in the mountains of northwestern Greece during the summer grazing. In the winter the flocks moved down to the plains of Thessaly. The dogs were known as aggressive protectors, only keeping their distance from strangers when the shepherds were present. At a very early age the puppies would begin to accompany the older dogs out with the sheep. The dogs lived roughly, subsisting on bread made from cracked wheat and whey as well as any small animals they caught. The numbers of good working dogs fell as the size of flocks decreased. Today the numbers of sheep are much smaller, although the dogs are still present in the mountains — and local travel guides suggest methods of dealing with

them. Recently conservation groups have been placing working pups with shepherds in areas of rising populations of wolves and bears.

The Greek sheepdogs are clearly related to the livestock guard dogs found in neighboring Albania, Macedonia, and Bulgaria. Within Greece, the effort to recognize the native dog breeds has struggled to compete with imported European breeds. There is also some effort to recognize the various types of livestock guardians found within Greece as separate breeds. The situation is uncertain at present and the dogs have not been exported outside of Greece in any significant numbers.

APPEARANCE

The Greek Sheepdog has been proposed for recognition by the FCI and is shown locally in Greece as the Hellenikos Poimenikos under a broad standard proposed by the Athens Canine Society. This standard describes a medium to large dog, both short- and long-haired, in a wide variety of colors and color patterns.

The Greek shepherds' traditional practice of cropping one ear continues today. The shepherds believed that the dogs would sleep on the uncut ear and, therefore, hear better with the cut ear. At times tails are docked as well.

CHARACTER/TEMPERAMENT

In general, the Greek Sheepdog is not suited for a companion dog, and he retains his sharp nature and strong sense of protection. In recent times some dogs have been brought to urban areas to work as home and business protectors.

Bulgaria

KARAKACHAN DOG
CAR-uh-kuh-chan
(Bulgarian Shepherd Dog)

Livestock predation is a tremendous problem in Bulgaria, which has one of the highest numbers and densities of wolves and bears in Europe, in addition to having golden jackals and lynx. Wildlife societies, carnivore conservations groups, and the Bulgarian Biodiversity Preservation Society Semperviva have all worked to supply livestock owners with Karakachan Dogs since 1997. The dogs are viewed as very successful livestock guardians, and their owners are obligated to disperse their puppies to more shepherds. In this way the tradition of using large, powerful livestock guard dogs is being restored after dark days for the people and dogs of Bulgaria.

Bulgarian Shepherd Dogs are related to the larger group of livestock guard dogs in the surrounding areas of Turkey, Greece, Macedonia, Serbia, and Romania. For many hundreds of years the Bulgarian shepherds used these powerful dogs to guard against predators as they moved their flocks between summer and winter grazing.

After World War II the Communist government nationalized the flocks, farms, and even mountain sheep enclosures. The native predators were hunted and poisoned. The government was hostile toward livestock guardian dogs and conducted massive exterminations several times over the next twenty years. At times the dogs' skins were marketed to the fur industry. In the end, livestock guard dogs were to be found only at some government farms and in extremely isolated rural areas.

In the 1990s, along with the first free elections, national pride turned toward saving the native breeds of sheep, horse, and dog. Breeders began to search out traditional Bulgarian livestock guardians. The

Karakachan Dog

traditional national breed has become an important symbol for the country, owned by both politicians and sheepherders alike. Many dogs are now being raised as companions, away from livestock. Semperviva, however, actively raises the traditional Karakachan Dogs to be placed with shepherds and cooperatively grazed flocks.

Unfortunately, there has been some crossbreeding with Caucasian Shepherds, Newfoundlands, and Saint Bernards, and many of the currently working livestock guard dogs are crossbred. However, the purebred nucleus is the Karakachan Dog, which is proving itself extremely successful against predators, outperforming the mixed-breed working dogs. Crossbreeding was accomplished in order to increase size and aggressiveness.

The Karakachan people, nomads of the Balkans, used their own livestock guardian, the Karakachansko Kuche. Being a minority, the Karakachan were especially persecuted under the Bulgarian Communist regime, and most of their herds were confiscated. Though for the most part the Karakachansko Kuche was assimilated into Bulgarian Shepherd Dog stock, it is still found in small but increasing numbers in Bulgaria and Macedonia. The Karakachan Dog or Bulgarian Shephard is recognized and registered by the Bulgarian government.

APPEARANCE

The Karakachan Dog is a rectangular breed that exhibits a variety of types, size, and coat. Shepherds in some regions tend to prefer dogs with heavy or rough heads and large muzzles. In other regions the dogs are leaner and trimmer. Both short- and long-haired types are acceptable. The dog has a double coat seen in several whole colors and bicolor patterns, although white dogs with dark spots are now quite popular. In the countryside shepherds often follow the old tradition of cropping one ear in the belief that the dog will hear better. Heights vary in males from 25 to 29 inches, with females slightly smaller. Weights range from 88 to 110 pounds in males, with females weighing slightly less.

CHARACTER/TEMPERAMENT

Sider and Atilia Sedechev, who are deeply involved in the Semperviva livestock guard dog project, report that in traditional Bulgarian sheepherding practices the Karakachan Dog behaves in a manner not generally seen elsewhere. The sheep are taken onto free pastures during the day with the shepherds. Some sheep spend the night outside, while others are penned. The dogs work in teams of two or three. If strangers approach the flock while on pasture, the dogs become visibly aggressive, although they will remain calm when passing through villages with the flock.

If they detect a wolf, they will leave the flock and chase it as far as two kilometers. Researchers have observed that after such encounters, wolves will leave that flock alone and attempt to raid other flocks. The shepherds believe that their dogs must show this level of dedication to harassing and even attacking wolves in order to combat the strong predator pressure in the area; therefore, they value physically strong, confident dogs. There are very few negative interactions between the public and the guarding dogs since locals appreciate the dogs' important work and know not to approach or threaten the flocks in any way. The only persistant problem is that hunters will shoot livestock guard dogs that attack their free-roaming dogs.

Several Karakachan Dogs are now working on farms in the United States. Their owners have found them to be very good with stock. They appear to be less aggressive toward people than many other breeds and quite levelheaded, with a steady temperament. Since they have been successful, more imports will probably occur in the future. There is an informal organization for the breed in North America that keeps track of the dogs and their offsrping.

Romania

BUKOVINA SHEPHERD
boo-ka-VEE-na
(Ciob nesc Românesc de Bucovina)

Agriculture is the most important economic activity in Romania, and it is home to very healthy populations of wolves, bears, and lynx. These predators cause significant economic loss to livestock every year. Romania is also home to an unbroken history of use of livestock guard dogs. Half of its sheep and cattle are still taken up into the Carpathian and Transylvanian mountains each summer. Eighty percent of these mountains are still covered in forest, which increases the predator pressure enormously.

Livestock guard dogs accompany all of the animals into the mountains. Many small, privately owned flocks are combined into groups of three hundred to five hundred sheep along with smaller numbers of cattle. Often the livestock owners hire a manager and his workers to accompany the flocks. The workers are also engaged in making cheese.

Some of the dogs are socialized to the sheep and live with them all winter, while others are owned by the workers and kept confined away from sheep during the off time. These two groups of dogs do not always work well together. Young dogs are placed with the flocks and expected to learn from the older dogs. Some of the sheep are penned at night, but others are allowed to graze all night close to the camps.

Romania faces several challenges in dealing with its predator problem. The country has slowly emerged from the draconian repression of Nicolae Ceau escu, who ruled for three decades. The rural people are desperately poor, and millions of dogs were abandoned when families were forcibly removed from the countryside, so that the country suffers from a huge stray dog problem. These dogs have interbred with the traditional guardian breeds, reducing the effectiveness of the working dogs. The working dogs are poorly fed on boiled corn flour and whey. Many dogs leave the flocks to hunt for food on their own. Significant numbers of dogs are killed by wolves, both on summer pastures and down in the winter villages.

The Carpathian Large Carnivore Project has studied these issues extensively. It is promoting the support of livestock guard dogs, including adequate

vaccination programs and the reduction of stray dogs in the country. One piece of good news is the renewed support for the three traditional Romania livestock guard dog breeds found both in the country and in the international arena. Although the Bukovina (pronounced *boo-ca-vee-na*) is recognized in Romania, it still lacks FCI breed status. It is the rarest and the largest of the Romanian livestock guards.

APPEARANCE

Obviously related to the other Balkan dogs including the Karakachan Dog, the Tornjak, and the Hellenikos Poimenikos, the Bukovina shows his mastiff influence in the size of his head, muzzle, and lips, his strong neck, and his massive build. Dogs average 30 inches in height, with corresponding weights. The double coat is abundant and moderately long, forming a ruff on the neck. The tail is bushy. The Bukovina is generally white with clearly defined patches of black, gray, or brindle, although solid white or black dogs are also accepted.

CHARACTER/TEMPERAMENT

The Bukovina is very active at night, patrolling his flocks or territory. He is aggressive towards predators or threats and suspicious of strangers, and he signals alarms with a loud bark. With his family and friends, he is playful and affectionate.

CARPATHIAN SHEPHERD
car-PAA-thi-an
(Ciob nesc Carpatin)

Recently recognized by the FCI, the Carpathian Shepherd was first described by Romanian authorities in the 1930s, when the Zoo-Veterinary Institute in Bucharest first began to describe and research the national shepherd dogs. It is related to the Sarplaninac and the traditional Karakachan Dog. Less

popular than the Mioritic, the Carpathian is still common in rural areas. He is a hardworking livestock guardian, although he is now finding a home in the cities as well.

APPEARANCE

A rectangular dog, the Carpathian is built more lightly than many other Balkan mountain dogs. He stands about 26 inches tall and weighs up to 110 pounds, without heavy mastiff influence. The standard has preferred fawn dogs with wolf gray markings since the 1930s, although other varieties are seen as well, including brindles, sables, and tricolored dogs with white markings. The Carpathian has a double coat with moderately long hair that is most profuse on the tail and neck.

CHARACTER/TEMPERAMENT

Many legendary stories relate the courage of Carpathians working in the mountains to protect their flocks of sheep against bears, wolves, and even wild pigs. The breed is still selected primarily for working ability. He is also known for his immense loyalty to master and flock.

MIORITIC SHEEPDOG
MEE-oh-ree-tic
(Ciob nesc Românesc Mioritic)

This is the most popular of the Romanian breeds, recognized by the FCI and owned by many public figures. Although the Mioritic is still primarily a working dog, its popularity has taken it into cities as a family and home guardian. Several Mioritic Sheepdogs have been exported to the United States, including the Romanian Embassy's watchdog in Washington DC.

APPEARANCE

The Mioritic is a bearded, long-coated dog that projects great size. Males stand 27 to 29 inches tall, with females slightly smaller. The coat is thick and requires some regular grooming. The color is white, light cream, or light gray, often with patches of black or gray.

CHARACTER/TEMPERAMENT

Affectionate toward his owner and good with children, the Mioritic is still a flock guardian that is suspicious of strangers and other dogs. He is smart and trainable but requires socialization and exercise if he lives in an urban environment. Without careful training and handling, some dogs can be either extremely aggressive or shy.

Turkey

AKBASH DOG
ah-k-bah-sh
(Akbas)

Evliya Celebi (1614–1682) explained in his massive *Book of Travels* that the Ottomans of the seventeenth century had two kinds of livestock guard dogs: *karabas* (black heads) and *akbas* (white heads). Today these two are known as the Kangal and the Akbash. Preliminary research on Turkish breeds by Dr. Peter Savolainen supports the assertion by many Turks that the Akbash and Kangal are two separate breeds and not types of the generic *coban kopegi* or shepherd's dogs. Akbash mitochrondrial DNA has haplotypes similar to those found in other dogs of this area, including the Saluki, Canaan, and Iranian dogs, meaning that they are closely related. However, the Akbash haplotypes differ more from those of the Kangal Dog, revealing that the two breeds have been genetically isolated from each other for a longer time. The two populations show little mixing, according to Savolainen.

Remarkably, the ancient Akbash (pronounced *ah-k-bah-sh*) has become a successful livestock guard dog here in North America, while its numbers have diminished in its native Turkey. The Akbash Dog is actually not well known in its homeland, an interesting reversal from the situation in North America, where the Akbash Dog is used in large numbers by ranchers in the West and has been the subject of livestock guard dog studies by the USDA.

In Turkey the Akbash Dog protects flocks against wolves, foxes, jackals, wild boar, and roaming dogs. When the flocks are not on pasture, the Akbash Dog must live companionably in the villages. In Turkey 30 years ago the Akbash Dog tended to be found in the area of Ankara and to the west. As society has changed and become more mobile, and sheepherding has declined, the Akbash Dog population has declined.

Small numbers of Akbash Dogs have been exported to Europe and elsewhere; about 50 have arrived in North America since 1978. However, there is no formal recognition of the breed except in the United States, where the breed is recognized by the UKC and ARBA. The Akbash population is now registered at more than two thousand, although many more dogs are unregistered working animals. Due to the breed's rarity in Turkey, imports are extremely difficult to obtain. Rescue dogs may be available occasionally.

APPEARANCE

The Akbash Dog is often described as elegant or graceful. He is generally leaner than the Kangal Dog and longer in the leg. Males weigh 90 to 130 pounds and stand 28 to 32 inches tall. Females are smaller at 75 to 100 pounds and 27 to 30 inches, and they are more feminine in appearance. Dogs bred and raised in North America tend to be larger due to the better nutrition and veterinary care they receive, as well as continued selection for increased size. The Akbash's appearance illustrates both his sight hound and his mastiff origins. Some dogs show more sight hound influence, with a narrower head, longer face, longer legs, less depth through the chest, and a more pronounced tuck-up. Other dogs show more mastiff influence, with a broader head and muzzle, more dewlap, heavier bone structure, and a broader chest. All dogs sport a long tail that curls over the back when they are alert.

The coat color is white, although there may be some light biscuit color on the ears, the top of the back, or the undercoat, but large areas of spotting or any other color are not acceptable. Under the coat, the skin is mottled with both dark and light skin; dark is preferable because of the sunburn risk with light skin. The nose, eyelids, and mouth are dark in color, with eyes colored light gold to dark brown. In Turkey the ears may be cropped. The Akbash Dog is double coated and sheds twice a year. Coats vary in length between smooth and short and long and thick. The Akbash Dog seems to be more heat tolerant than many other livestock guard dog breeds. The shorter-haired coats are well suited to hotter climates such as the southern coast of Turkey and some areas of North America as well.

CHARACTER/TEMPERAMENT

The Akbash Dog has the size, power, and protective nature of the mastiff combined with the speed of the sight hound. Females and males guard equally well. They are early to sound the alarm, with excellent senses of hearing and sight. Like all livestock guard dogs, they are independent and territorial. They are wary of people they do not know and will not accept intruders on their own property, especially when their owners are absent. However, this does not mean that Akbash Dogs are vicious with strangers off their property; they should not be. Studies in the United States show the Akbash Dog to be one of the best livestock guard dog breeds, combining strong response to predators and roaming dogs with reduced aggression to people. Akbash Dogs are working successfully in both fenced pastures and rangelands protecting many different species against many predators, including the fiercest.

ON THE FARM

Robyn and Jim Poyner
Poyner Goat Company, Monett, Missouri

Robyn and Jim Poyner raise Boer goats and Boer-Spanish crosses that they sell for meat; Angus steers and poultry are also part of the farm's operation. In their area of southwestern Missouri, the Poyners face predation from coyotes, packs of roaming dogs, bobcats, and the occasional mountain lion. They have experienced situations of human theft as well. All of these problems ended with the introduction of livestock guard dogs about ten years ago. Today they own ten livestock guard dogs from several breeds.

The Poyners work their dogs in packs, believing that this is the most effective tactic for their situation. Robyn observes, "One dog, in most cases, cannot protect the stock and still fend off a pack of dogs or coyotes, and multiple dogs are needed to hold off a truly hungry large cat."

Managing multiple dogs requires a special effort. Robyn advises, "Get to know each dog you own. Understanding the dog's pack status and mentality will help you to understand things that the dog does in the field and how it acts or reacts toward other dogs." Robyn relates that their greatest challenge in working with livestock guard dogs over the many years has been "learning to think like the dogs and understanding what they were doing."

The Poyners had an important lesson in the value of this approach. Robyn remembers, "My husband called me at work one day to tell me that one of our dogs had been found with a dead kid goat in her mouth. He was ready to put the dog down. I asked him if he had seen the dog maul or kill the kid and his answer was no. I asked him to inspect the dead kid. He called me back a short time later to tell me that the kid appeared to have been premature and stillborn. We believe the dog was getting rid of the remains that would only draw in a predator if left in the pasture."

The Poyners often get calls from owners who are upset that their dogs are sleeping all day and not staying close to the stock. Robyn notes that "upon questioning the owners about their setup, predation situation, and so on, we find that the dogs are working all night and resting during the day, when there is little or no predator activity: proof that even if they cannot detect it, the dogs are serving the purpose." Robin shares her knowledge of working dogs with others on a livestock guard dog e-mail list (see Resources, page 216).

The Poyners believe that one key to success with livestock guard dogs is to train them early to respect the fences. The Poyners use electric wire to teach puppies and young dogs to avoid all fences. They also feel strongly that all of their dogs must be safe around children, so puppies are deliberately socialized to help them develop a positive attitude toward children.

The Poyners are unusual in that they have deliberately set out to gather experience with several breeds of livestock guard dogs. Including their own dogs and rescues, they have worked with Akbash Dogs, Anatolians, Central Asian Shepherd Dogs, Estrela Mountain Dogs, Gamprs, Great Pyrenees, and Karakachans. Robyn's advice to a potential owner is "to find a breed that suits your situation and needs. Find a breed that appeals to you, not just another dog. Also, understand that livestock guardian dogs range from mellow (unless challenged) to very tough temperaments. Evaluate your own experience and abilities and choose a breed that fits you. Find people who have real experience with the dogs and ask questions. You should never be afraid to ask questions."

The breed standard in the United States notes the extreme maternal nature of Akbash Dogs. They form early and strong attachments to their charges. Typically Akbash puppies are raised with stock from an early age, and breeders will recommend this practice if your dog is to work as a livestock guard dog. This maternal instinct can be seen in the dogs' submissive behavior toward livestock, and especially toward young animals. Akbash Dogs also tend to be affectionate toward their family members, although they still require socialization and basic obedience work.

SPECIAL CONSIDERATIONS

The Akbash Dog is physically slow to mature, and males may not reach full size until the age of three or more. Akbash puppies can be very active, and they are great chewers; breeders suggest you guard against mouthiness in puppies. Adults can be great diggers — or excavators, as some breeders note. They will also wander unless they are well contained. Adults can be aggressive toward other dogs, even off their property. If you own other dogs, be aware that Akbash Dogs are more likely to get along with smaller, nondominant dogs of the opposite sex.

Akbash Dogs are excellent livestock guardians as well as home and estate guardians. Some breeders do not believe that Akbash Dogs are suited to city or urban life and will not sell puppies to those who live in such areas, believing that an Akbash in this situation is likely to end up in a rescue program. Others believe that an experienced dog owner who is committed to constant socialization and control can enjoy the Akbash as a companion dog in the city.

Due to the strict OFA breeding requirements of Akbash Dogs International, as well as their tightly muscled conformation, purebred Akbash Dogs may have the lowest incidence of hip dysplasia among the North American livestock guard dog breeds. There are, however, some incidences of osteosarcoma, ACL injuries, seizures, and cardiomyopathy.

KANGAL DOG
KAHN-gall
(Kangal Kopegi)

"Beyond Kayadibi the country dogs were the largest and most savage of any I had met. In build they were like Newfoundlands, but larger, with black head or muzzle, yellow body, and long curling tail. From nearly every flock that fed within a half-mile of the road a dog would presently detach itself and come lumbering across country to the attack."
—*W. J. Childs,* Across Asia Minor on Foot *(London, 1917)*

Childs was certainly not a dog lover, but he accurately described the native Kangals (pronounced *kahn-galls*) he met along the old Silk Road, outside of Sivas in Turkey. Sheepherding has been a way of life in this region for hundreds of years, and references to the shepherds' "great yellow dogs with black faces" go back to the fourteenth century. The principal predator here is the wolf, but other predators include brown bears, feral dogs, mountain cats, foxes, jackals, and wild boars.

The Kangal Dog has been declared both the national dog of Turkey and a national treasure, celebrated at an annual Kangal Dog Festival and regular Kangal Dog symposiums. The Kangal Dog

is conserved at various Turkish military and university facilities, including two breeding centers in Sivas. It is technically illegal for non-nationals to export Kangals from the province. In addition, the United Nations has funded small grants for projects that demonstrate and encourage the use of Kangal Dogs in conservation efforts to protect wildlife.

In Turkey the Kangal Dog works actively with shepherds who take their flocks out for grazing during the day and return to the villages at night. In the villages Kangal Dogs are expected to be gentle with children and tolerant of neighbors. They are not allowed to run free but are confined outside the home or with the sheep. They are fed barley mash, scraps, and bones. During the summer flocks often make the journey to high summer pastures known as *yaylas*, far from roads and people. Two or three dogs will accompany flocks of two hundred to three hundred sheep. Shepherds spend the summer in isolated huts, working with the dogs to guard the sheep from wolves that often prey at night. Kangal Dogs are known for their fierce battles with predators, first intimidating them through barking, but pursuing them when necessary. Puppies grow up in the village until they are old enough to accompany the older dogs and learn from both them and the shepherds.

In Turkey the Kangal Dog is recognized by the Federation of Dog Breeding and Cynology, operating in association with the Turkish Ministry of Agriculture. The federation is seeking FCI recognition for the breed. The Kangal Preservation Fund works to vaccinate Kangal Dogs in and around Sivas. Kangal Dogs were first imported to North America about twenty years ago, and the breed is recognized by the UKC. In Britain, where Kangal Dogs have been bred for nearly forty years, the Anatolian Karabash Dog Club maintains pedigree records of pure Kangal Dogs.

The Kangal Dog is still rare in North America, but the number of breeders is growing, and it is possible to obtain a pup or young adult after a short wait. Kangal Dogs are increasingly being raised on farms as working livestock guardians.

APPEARANCE

The Kangal Dog is large, with heavy bone structure, and projects a powerful image, though he is also fast and agile. He has been described as being lionlike or as a natural mastiff, without extreme head size, dewlaps, loose lips, head wrinkles, and large bulk. When the dog is alert, the tail is carried in a curl over the back, although it may be carried low at other times. The Kangal males achieve a height of 30 to 31 inches and weigh 110 to 145 pounds, with females slightly smaller. The natural variety in a landrace breed is evident in the range of the breed's appearance, from being more to less mastiff-like. Females are often more feminine in their appearance.

The Kangal Dog coat handles extremes of weather, with a natural short coat in the heat of summer and a dense double coat in winter. Kangals do shed twice a year. They are generally light dun to gold in color, though depending on the amount of dark gray or black in the outer guard hairs, some Kangals can appear almost sable. A black mask may completely cover the muzzle and extend over the top of the head. The pendant ears also carry black color. In Turkey the ears are often cropped, but not always. In North America cropping is permitted only on imported dogs. Small amounts of white are allowed on the feet, chest, or chin, and white blazes on the chest may be outlined with dark hair. The tip of the curled tail is often black, as is a spot on the middle of the tail. The presence of other solid colors or patterns is evidence of non-Kangal breeding.

CHARACTER/TEMPERAMENT

Temperament was of prime importance in Turkish villages in the past, and aggression toward humans or stock was never tolerated. As a result the Kangal Dog is noted for his solid temperament and gentle manner with livestock, children, and pets. Compared with other livestock guard dog breeds, Kangal Dogs tend to be more people-oriented and less standoffish. Females are noted for being more affectionate than males. The Kangal is often a great judge of character and will accept visitors or workers at his home much more reliably than other livestock guard dog breeds. Owners describe Kangals as gracious with visitors. People who own other livestock guard dog breeds and Kangal Dogs often describe their Kangals as being more intelligent and clever.

When Kangal Dogs hear or sense disturbances in the distance, they bark in response. They prefer to intimidate predators that threaten their flock or family, but they will attack animal intruders. At first they will place themselves between the threat and their stock. If their warning barks are ignored, they will confront the predator with a roar and an attack if necessary. An adult Kangal will throw his shoulder against a wolf to knock it down and then attack the throat and hind legs. Kangal owners see these same behaviors practiced in play.

Kangal Dogs grow up slowly, even compared to other livestock guard dog breeds. At around two to two and a half years of age, Kangal Dogs suddenly leave their puppyhoods behind and become more serious, protective, and watchful. Although many Kangals that are raised with stock from puppyhood are reliable at an early age, others need your patience until they mature. They often take a remote or high viewing spot to watch their stock, although they will make regular patrols around their territory. Kangal Dogs tend to be calm and even placid during the day and more active at night, both patrolling and barking.

Kangal Dogs are territorial and they behave quite differently at home than when they are away from home. Like many livestock guard dogs, adult Kangal Dogs can be dog aggressive, especially with dogs that violate their space or sense of self. They will kill small predators that invade the farm or pasture, and these may include small pets they do not know.

All of these breed traits make the Kangal Dog well suited to life as a farm or family guardian in addition to a livestock guard dog. Indeed, in Turkey Kangals are popular home guardians in the city, where they guard yards and gardens and play affectionately with their families. Kangal Dogs also live in many family situations in North America, but as with all livestock guard dog breeds they require socialization and consistent training to be good canine citizens.

SPECIAL CONSIDERATIONS

Kangal Dogs can be goofy and adorable as puppies but are also famous as powerful chewers and diggers. They will also roam large distances if given the opportunity.

As with all livestock guard dog breeds, hip dysplasia is a potential problem, and breeding stock should be screened for it. ACL injuries have also been reported. Bloat is not generally a problem with this breed.

KARS DOG

Northeast Turkey shares a common livestock guard group with its neighbors Armenia and Georgia. These Turkish dogs, which recently received the name Kars out of respect for the central region of that area, are closely related to the ovcharka group of the Caucasian region (see page 153). Kars Dogs are used for both flock and home protection. Some Kars stock has been crossed with ovcharkas and other large mastiff breeds for dog fighting stock.

A number of Kars Dogs have been imported to North America and Europe as part of Anatolian

breeding stock. They do not have individual recognition yet, although Turkish authorities and owners are beginning to research and record Kars breed information.

APPEARANCE

Kars Dogs are powerful and strongly muscled, with large heads showing mastiff influence. Coats and colors vary, although owners often prefer the wolf-gray, long-coated dogs. Ears are often cropped. The average height is 28 inches, although taller dogs exist.

CHARACTER/TEMPERAMENT

Many of these dogs are more aggressive and territorial than either Kangal or Akbash Dogs. They are dogs for serious owners in rural areas.

ANATOLIAN SHEPHERD DOG
an-ah-TOLL-ee-an

Anatolian breeders in England and North America believe that all the historic livestock guard dogs of Turkey are part of one greater landrace group, which they have named the Anatolian Shepherd Dog. In modern Turkey you can find a wide range of shepherd's dogs, from the fawn-colored, black-masked Kangal to the white Akbash, pinto or other colored dogs, large mastiff-type dogs increasingly referred to as Aksaray Dogs, rough-coated and multicolored ovcharkas, and Kars Dogs. No doubt, there have long been excellent *coban kopegi* (shepherd's dogs) of varied types, just as there have been centers of more selective breeding to a certain type. Many of the *coban kopegi* resemble dogs found in a huge area from Iraq to Afghanistan. According to many observers in Turkey, modern transportation has increased the presence of the *coban kopegi*, just as modern society has decreased the numbers of shepherds.

At this time, the *coban kopegi* is not a recognized breed in its native Turkey, and the name Anatolian Shepherd is a foreign invention. Paradoxically, the Anatolian Shepherd is one of the most numerous and successful livestock guard dog breeds in North America. How did this situation occur?

A British archeologist exported two Kangal dogs to Britain in 1965. Since she was told they were called *karabash* ("black head"), she helped establish the Anatolian Karabash Dog Club in Britain in 1971. (*Anatolia* has long been used as a name for the historically greater geographic region of Turkey.) The standard for this breed specified a short-coated, fawn-colored, black-masked dog. After additional imports of both Akbash and *coban kopegi* by other

people, a new breed club was established known as the Anatolian Shepherd Club of Great Britain. This new group promoted the Anatolian Shepherd Dog in all colors and coat types.

A similar club was founded in the United States. The Anatolian Shepherd Dog Club of America also accepted all colors and coat types. In the last few years fanciers have been breeding more dogs that resemble the Kangal type, as they refer to it, and it is this image that is being used more and more to represent the Anatolian Shepherd Dog.

In North America the Anatolian Shepherd Dog quickly proved itself to livestock producers, and its population increased rapidly. Dr. Ray Coppinger was an early importer and promoter of Anatolians as well. Though in Turkey the *coban kopegi* work with the shepherd and return to the villages with the flocks, in North America they are often asked to be alone with the flocks.

Today the Anatolian Shepherd breed draws from many types of dogs found in Turkey. Due to this diverse gene pool, the Anatolian has a greater variability in both appearance and behavior than some other livestock guard dog breeds. Much of the information contained in the Akbash, Kangal, and other breed entries may be applicable to your particular Anatolian, although color alone is certainly not an indicator of breed or breed characteristics.

At present there are many Anatolians and part-Anatolians in shelter and rescue situations in certain areas of the country. This is probably due to a combination of uninformed or unprepared owners and poorly bred, unregistered Anatolians or Anatolian crosses. This cannot be repeated too often: you get what you pay for. Also, an Anatolian is a challenging dog that demands a devoted owner. He needs a job to do or he will create his own. Finally, an Anatolian must be socialized, whether he will be working as a full-time guardian or as a companion.

APPEARANCE

The Anatolian Shepherd is a large, powerful dog. He should appear slightly longer than tall. Males stand at least 29 inches tall and weigh 110 to 150 pounds. Females are slightly smaller, at 80 to 120 pounds and heights beginning at 27 inches. The Anatolian should not be too mastiff-like, nor should he lack substance and appear like a sight hound. All colors and colors patterns are accepted with or without a black mask, including pinto, brindle, and white "Dutch" markings (a white collar and blaze with white legs, like a Dutch rabbit). Coat length can vary between one and four inches. All Anatolians are double coated and shed twice yearly. All tail carriages are allowed. Cropped ears are often seen in Turkey but are not acceptable on dogs bred in the United States. The skull is broader in males than in females, with a slight centerline furrow. The eyes are almond shaped and dark brown to light amber in color.

CHARACTER/TEMPERAMENT

The mature Anatolian should be a calm and observant dog. His guarding technique may vary from staying close to his flock to removing himself to a distant observation site. He will bark to deter potential predators and in response to disturbances. Barking will escalate to more threatening intimidation if necessary. Puppies can be great diggers and chewers. If not well fenced they will develop the habit of roaming.

Anatolians can be reserved with strangers when away from their property. At home they will be highly territorial and responsive to threats, although a well-trained and well-socialized dog should accept visitors. Well-socialized dogs should tolerate other dogs when they are away from home, but some will not accept strange dogs on their property. An Anatolian who is not raised appropriately may become overprotective, aggressive, and difficult to control.

SPECIAL CONSIDERATIONS

Health concerns include hip dysplasia and sensitivity to anesthesia. Bloat is not generally a problem with this breed. Entropion and thyroid deficiency occur in some lines.

Armenia

GAMPR
GAHM-peer
(Armenian Shepherd)

The Armenian Gampr belongs to the ovcharka group of dogs found throughout the Transcaucasian area (see also Caucasian Mountain Dog, page 153). There is no official standard or recognition for the breed at this time, although there is growing popular and governmental support for this historic dog of the mountains. The Gampr is described as a powerful and reliable livestock guard in Armenian literature dating back to the ninth century. Large numbers of dogs were exported to the Soviet Union, leaving the native dogs relatively scarce by the 1930s. About this time, Gamprs were also taken to Germany where they were first exhibited in Europe. The Armenian national government has now begun the process of procuring official FCI recognition and registration and developing a standard. Several other countries, including Azerbaijan, Georgia, Iran, and Turkey, also claim to be the true home of this breed. Several Gampr dogs have now been imported into the United States.

APPEARANCE

In Armenia, the Gampr is a large, strong dog with a powerful head. His ears and tail are often cropped. A wide variety of colors and color patterns is acceptable, since the Gampr is still bred to work and not to a standard. The Gampr comes in two varieties: short and long coated. The long-coated, heavier dog tends to be found in the mountainous areas, while the lighter, short-coated dog is usually found in lowlands. Heights range from 28 to 30 inches or even more in the mountain type.

CHARACTER/TEMPERAMENT

The Armenian Gampr is a calm, serious yet curious dog, which will respond to challenges quickly and with great power. He requires an experienced owner who understands that although his dog is loyal, he is an independent thinker and not blindly submissive.

Eastern Europe

SOUTH RUSSIAN OVCHARKA
of-SHAR-ka
(South Russian Shepherd Dog, Youzhak)

This bearded livestock guardian was developed in the Crimea in present-day Ukraine at least by the early nineteenth century. The breed may be related to the other bearded guardians of the greater area such as the Mioritic, or to others introduced with Merino sheep from Spain. Early in the twentieth century the reduction of wolves in the area and the Russian revolution combined to severely reduce its numbers. The first attempt at a revival of the breed was conducted at an official breeding center in Crimea, and dogs were later exported to both Europe and North America. World War II devastated these efforts. After the war, it was necessary to use Komondorok and other dogs to revive the breed once again. The breed has achieved some popularity in Russia and has been used both as a military guard and as a property watchdog. Since it was accepted by the FCI the South Russian Ovcharka (*of SHAR ka*) has been raised in Europe, primarily for showing. Since the 1990s it has been easier to export Russian dogs, and South Russian Ovcharkas are now seen occasionally in North America.

APPEARANCE

The South Russian Ovcharka must stand a minimum of 24 inches in height. He has a broad head and his ears are not cropped. He is rectangular in shape and fairly long in the leg. The tail is held low and raised to level when the dog is alert. The double coat is thick, coarse, and long. The color is white, gray, or tan; a white coat may or may not have gray or tan spots.

CHARACTER/TEMPERAMENT

Unfortunately, many of these dogs have proved difficult for some owners to handle, although other owners claim that with vigilant socialization the South Russian Ovcharka can be well behaved. The breed is very dominant and independent by nature. No South Russian Ovcharkas appear to be working in North America as livestock guardians at present.

SPECIAL CONSIDERATIONS

If the dog's coat is not groomed it will mat, and the beard can become stained or smelly if not cleaned regularly. Owners in North America report that their dogs are prone to hot spots and skin irritations in hot or humid climates.

CAUCASIAN MOUNTAIN DOG/ CAUCASIAN OVCHARKA
of-SHAR-ka
(Caucasian Shepherd, Kavkazskaïa Ovtcharka)

There is a common livestock guardian group found throughout the Caucasus Mountains from eastern Turkey to Georgia, Armenia, Azerbaijan, and Iran in the Transcaucasus grasslands. This group of dogs is now commonly called the *ovcharka* group; *ovcharka* is a Cyrillic word meaning "shepherd," "shepherd's dog," or "sheepdog." There is some rivalry among countries in the area as to which is the true home of the ovcharkas. Different areas in the region have their own names for these dogs and long-established landrace breeds developed from the ovcharka type. Some are seeking recognition for their own particular types, such as the Georgian Nagazi.

Caucasian Mountain Dogs have long been used as both livestock and property guards. Recognition of this specific breed began in the 1930s, when it was shown in Germany. Dogs were also taken from the Caucasus into Russia in large numbers to be bred by the Soviet Red Army and state-run kennels as aggressive guard and patrol dogs. The dogs were often crossbred with other breeds, including Saint Bernards, Central Asian Shepherds, German Shepherds, Sarplaninacs, and others. The Moscow Watchdog was developed using Caucasian Mountain Dogs and Saint Bernards.

By the 1960s the Caucasian Ovcharka was a popular breed with a registry and regular shows. The breakup of the Soviet Union in the 1990s led to the closing of many kennels and the abandonment of many dogs that were no longer needed by the military or as factory guards. Following this disruption, the FCI recognized the breed as Russian and the Russian Kynological Federation as its national club and registry. Since an estimated 70 to 80 percent of the breed's population is now believed to be crossbred, the FCI requires the individual inspection of all Caucasian Ovcharkas before registration. Although there are different types and colors within the breed, the RKF standard prefers the bearlike Georgian type in wolf-gray colors.

The Caucasian Ovcharka is also a companion and show dog in Europe and North America. The breed is recognized by the UKC as the Caucasian Ovcharka and by the AKC Foundation Stock Service as the Caucasian Mountain Dog.

APPEARANCE

The Caucasian Ovcharka is a powerfully built dog with an imposing presence. The wedge-shaped head is an important feature for this breed, and it has deep-set, slanted eyes, a broad skull, and a blunt muzzle. Typically the Caucasian Ovcharka has a medium-length to long double coat. The ears are traditionally cropped close to the head, except where cropping is forbidden in Europe. The absence of earflaps combined with the long hair found on the cheeks, back of the skull, and neck form a striking mane that gives the Caucasian Ovcharka his distinctive bearlike appearance. The legs and tail are heavily feathered, and the feet are furred between the toes. This breed requires weekly grooming and sheds heavily.

The coat may have wide color and pattern variety, except red and white coloring similar to that of the Saint Bernard, solid black or brown, or black and tan. Although the minimum height for males in the United States is 25½ inches, many breeders prefer much larger dogs. Dogs weigh 100 to 150 pounds or more. The tail may be carried low or curled over the back.

CHARACTER/TEMPERAMENT

The Caucasian Ovcharka is a highly territorial dog that is often dog aggressive. He requires careful and continuous socialization throughout his lifetime by an experienced and committed owner. Although he will strongly bond with his owner, he might not obey the family's children and may overreact to protect them in play. The Caucasian Ovcharka can react very quickly when he becomes protective. Experienced breeders recommend six-foot fencing. Caucasian Ovcharkas are also noted den diggers.

Although well-bred Caucasian Ovcharkas can and do work as livestock guard dogs, please be aware that strains of this breed were bred as aggressive property guards or fighting dogs. Most Caucasian Ovcharkas are companion or protection dogs. Many dogs are generations removed from their traditional work and do not exhibit a measured response to a threat. Caucasian Ovcharkas are more trainable and people oriented than many other livestock guard dogs due to their history and selection as guard dogs; however, this also makes them less suitable as remote or full-time guardians. Caucasian Ovcharkas are active at night, patrolling and barking.

SPECIAL CONSIDERATIONS

Due to their heavy coat, Caucasian Ovcharkas are not suitable for tropical climates. The health concerns for this breed include a high rate of elbow and hip dysplasia, which ethical breeders are combating through screening. Entropion, hypothyroidism, bloat, and food allergies occur at low rates.

Central Asia

CENTRAL ASIAN SHEPHERD DOG
(Central Asian Ovcharka, Sredneasiatskaia Ovtcharka)

The traditional homelands of the Central Asian Shepherd encompass mountains, high plateaus, vast deserts, and grassy steppes. The great nomadic tribes in the area eventually gave way to settled peoples and then collectivization and industrialization under the centralized rule of the Soviet Union in the early twentieth century. The newly independent republics — Kazakhstan, Kyrgyzstan, Tajikistan, Turkmenistan, and Uzbekistan — are struggling but also reasserting their native peoples' cultures.

The large dogs of this region were traditionally guardians of property, including caravans, livestock, and homes. They were known by a variety of local names and bred to specific needs. Under Soviet rule they were taken into government breeding kennels and drafted into service as guard dogs for factories and the military. They were recognized collec-

tively as a single Russian breed, the Central Asian Shepherd, by the FCI in 1993. The Central Asian Ovcharka remains a popular dog in Russia. In the 1990s it became possible to import these dogs from Russia or directly from central Asia into Europe and North America. The UKC and the AKC Foundation Stock Service both now recognize the Central Asian Shepherd Dog. In the future individual countries may move toward recognizing their own types of this breed as distinct breeds, such as the Alabai from Turkmenistan.

APPEARANCE

The Central Asian Shepherd is a massive and powerful dog. The average male stands 29 to 31 inches tall and weighs 120 to 145 pounds. Females are 25 to 29 inches tall and weigh 90 to 125 pounds. Extreme size or weight may be a sign of crossbreeding with larger breeds. The head is broad and large with a deep, blunt muzzle. The ears are normally cropped close to the head. The neck has a heavy dewlap. The tail may be docked short or kept natural.

The coat is thick but may vary from 1½ to 3 inches in length. The Central Asian Shepherd is seen in many colors and patterns. Dogs from central Asia are generally yellow, tan, brown, sable, brindle, gray with white markings, or masked in black. Dogs from the Russian sentry lines are typically black with or without markings.

CHARACTER/TEMPERAMENT

Dogs imported to North America come from various sources and therefore exhibit a variety of temperaments. Dogs from central Asia and working livestock origins are the most suited to jobs as farm and family guardians in a well-fenced area that has serious predator pressure. These dogs are highly territorial and will protect their property with appropriate measured responses. They need the guidance and companionship of their human owners and are not suited for use as remote guardians. Owners must commit to continuous socialization.

Dogs from Russian kennels that bred for sentry work are more trainable but may not exhibit good livestock guardian qualities. Beware of dogs from any kennels specializing in fighting dogs or show dogs that are tested against other dogs. Some modern lines of these dogs are being bred with undershot jaws to improve their biting grip and are no doubt crossed with other fighting breeds. Unfortunately, dog fighting has become widespread through many of the former Soviet-controlled republics in the last decade.

Selected from appropriate lines, Central Asian Shepherds are generally a more relaxed and people-oriented breed than its ovcharka cousins. They will accept strangers if properly introduced. They are generally fine with family pets if they are raised with them. Introduce new dogs carefully, since Central Asian Shepherds are dominant toward strange dogs. They are protective and territorial by nature. They are also nocturnal barkers, especially the females, who also have a high-pitched howl. They shed heavily once a year. Like most livestock guard dogs, they are diggers, and they require a 6-foot fence or livestock fence combined with a hot wire.

It is strongly suggested that you meet the parents of any puppy you are considering. Adopting adult rescue dogs is a job for experienced owners only.

SPECIAL CONSIDERATIONS

Health concerns, in order of common occurrence, include ACL injuries, entropion, and elbow, shoulder, and hip dysplasia. The Central Asian Shepherd is generally a healthy breed due to its broad diversity in types; however, it needs to be bred for working ability, health, and longevity rather than show-ring popularity or fighting prowess.

Afghanistan

SAGE KOOCHEE
sahg-i KOO-chee

For many centuries the nomadic Kochi people traveled with their flocks of sheep, goats, and camels between the Afghan highlands in summer and the lowlands of Afghanistan, Pakistan, and Iran in winter. Before the Russian invasion and the Taliban rule that followed, millions of Kochi people lived this traditional life. The Sage Koochee — "the dog of the Kochi"— guarded the caravans and flocks from wolves, big cats, and human thieves. There were many regional variations of the breed, including the Djence Palangi, the Djence Sheri, the Sage Rama, and the Sage Mazandarani.

The Russian Army destroyed much of the Afghan infrastructure, including the important irrigation and water systems, which plunged the country's agricultural system into crisis. This was followed by four years of heavy drought beginning in the late 1980s. Many refugees fled to surrounding countries. Both the Russians and the Taliban shot dogs on sight. The countryside was also heavily mined. Following the overthrow of the Taliban in 2001, the Food and Agriculture Organization (FAO) of the United Nations was able to enter the country and assess the state of agriculture. The organization estimated that up to 85 percent of the prewar livestock had been killed through warfare, drought, and mines. Large programs of vaccination followed, for both the livestock and the remaining dogs, since rabies had become rampant. (The FAO estimated that at the time of its arrival, rabid animals bit four hundred Afghans a month.) The country still faces a huge stray dog problem. Most of the Kochi people have settled in encampments on the edges of cities, since land ownership has taken away much of their traditional grazing land.

Koochee dogs have reemerged as a breed from the remote areas and neighboring countries. Unfortunately, the traditional Afghan love of dog fighting has also resumed, attracting thousands of spectators in Kabul during the fall and winter months. Afghan dogs are being crossed with dogs brought in by foreigners, both Western breeds and ovcharkas from Russia and other countries to the north, to breed better fighters. Though some Koochee dogs have been exported to Europe, there is no organizing body in Afghanistan for dogs, and the situation is quite chaotic at this time. It may prove very difficult to identify true Koochee types in the future.

APPEARANCE

The Sage Koochee ranges from very heavy ovcharka or mastiff types to lighter desert types. All colors and color patterns are present. Coats vary between short and medium-long. The ears and tail are often cropped.

CHARACTER/TEMPERAMENT

Traditional Koochee dogs are territorial and aggressive. They are generally not good family pets.

Tibet

KYIAPSO
kee-OP-so

The word *apso* in Tibetan means "hairy." Many small Apsos, as they are called in Tibet, were once found in temples and homes. There is a famous photo of the Dalai Lama fleeing from the Chinese in 1959 with a small Apso in his arms. Today the small Apsos are known as Lhasa Apsos in the West. The larger Dhokhi Apsos, which often worked as herders, are known as Tibetan Terriers in North America. When Westerners visited Tibet they also observed large Apsos guarding homes, temples, and the livestock of the nomads, particularly around the region of Mount Kailash on the western high plateau. These larger, bearded guardians have come to be known as KyiApsos ("hairy dogs").

In many ways the KyiApso combines the traits of the Tibetan Mastiff and the smaller Apsos. Some of the playfulness and charm of the smaller dogs is found in the KyiApso, along with the serious guardian qualities of the Tibetan Mastiff. Although the KyiApso has also been called the Bearded Mastiff, both the Tibetan KyiApso Registry and the American Tibetan Mastiff Association believe that their two breeds are definitely separate. The coats of these dogs differ, and the KyiApso is a more slender, more agile, and longer-legged dog. Nonetheless, there are many similarities between the dogs, and no doubt some interbreeding has occurred, both past and present. At times smooth-faced pups are produced in KyiApso litters, probably because of this shared heritage.

The KyiApso is very rare in his homeland, and fewer than one hundred dogs are thought to exist outside of Tibet. About forty KyiApsos can be found in the United States, although enthusiasts hope for more imports from Tibet.

APPEARANCE

With his full, long double coat and bearded face, the KyiApso appears larger than he really is. The outer guard hair is 3 to 6 inches in length. The hair on the face and muzzle is profuse. Any coat color is allowed, but the most common colors are black and tan, black and gold, black and silver, and black with a white chest spot. Multicolored coats also occur, and other colors found in Tibet include gold, white, mahogany, and chocolate brown. Under all of that hair, the dogs are actually medium in size, without overly heavy bone structure. Dogs born and raised in the West tend to be larger than dogs in Tibet due to improved nutrition, but they should not be taller than 28 inches or heavier than 100 pounds.

The tail is set high and carried in a full curl when the dog is alert, although it may be carried lower when the dog is relaxed. The lips are tight and never pendulous. The KyiApso moves with a characteristic bounce, and he is very fast and agile.

CHARACTER/TEMPERAMENT

Owners report that the KyiApso is very territorial and on his home ground should not be approached casually. Like the Tibetan Mastiff, the KyiApso should not be used as a full-time remote guardian

or left alone for long periods without the company of other dogs or people. Bored dogs can be extremely destructive. Barking problems need to be controlled early. Dogs raised without socialization, training, and a strong owner can be dangerous. However, KyiApsos can be laid-back companions for people who enjoy their sense of humor, provide them with exercise, and train them properly. KyiApsos also live happily in groups.

TIBETAN MASTIFF
(Do-Khyi)

The Tibetan Mastiff is not the ancestral form of all the mastiffs or the *molossers*, nor is it the mysterious dog lost in the mists of time in ancient Tibet. The reality is more complicated. *Tibetan Mastiff* is the Western name for a breed created from a wide group of related dogs found throughout the vast Himalayas, including Tibet, Nepal, Bhutan, northern India, and western China into Mongolia.

The historical records are clear: the large guard dogs of this type have been in this area throughout recorded time. They go by many names, and several were used in the creation of the standardized Do-Khyi, the traditional "gate dog"; Bhotia, the agile livestock shepherd; and Tsang-Khyi, the largest of the guard dogs.

In their native homeland these dogs were literally everywhere. Nomadic herders of sheep, goats, yaks, and horses needed protection from predators including the wolf, the black bear, the snow leopard, and the lynx. Caravans worried about not only the four-legged predators but the two-legged ones as well. In villages the dogs were tied at the doors of houses or kept in the walled courtyards. They were often turned loose at night to roam and guard the entire village. As reported by many visitors, the monasteries were home to the finest of these dogs.

The traditional Tibetan way of life began to come to an end in 1950 when the Chinese invaded. Following an uneasy agreement and an unsuccessful revolt, the Dalai Lama fled Tibet in 1959. The nomad areas were reformed into communes, the caravans ended, the monasteries were closed, and ethnic Chinese resettled the urban areas of Tibet. The commune system failed to feed the people, however, and individual ownership of animals was restored in 1981. The Chinese government has recently organized Tibetan Mastiff breeding centers and authorized private companies to encourage export of dogs from Tibet and China. However, the true state of well-bred Tibetan dogs is unknown.

Tibetan dogs occasionally made their way to the West but never in large numbers. A handful of dogs came to Britain early in the twentieth century, and a standard was written and approved in the 1930s under the name Tibetan Mastiff. President Eisenhower was surprised to receive two Tibetan Mastiffs from Nepal in 1958 after he had requested two much smaller Tibetan Terriers. In the 1970s several dogs came to the United States from Nepal, some imported directly and others as cover for drug smuggling. The founding population for the Tibetan Mastiff in Europe and North America remained quite small until the mid- to late 1990s, when new

breeding stock was imported. In 2004 the FCI recognized the Do-Khyi as a Tibetan native dog. To date, about three thousand of these dogs have been registered in the North America, and about three hundred and fifty reside in Britain.

APPEARANCE

The Tibetan Mastiff in North America shows variation in size, type, and temperament. He projects power through his heavy bone, abundant coat, and serious expression. His head is heavy, with a rounded skull; small, dark eyes; a broad, blunt muzzle, often with pendulous flews; and wrinkles at the eyes and mouth. Tibetan Mastiff males generally stand 25 to 28 inches tall, with females slightly smaller, but there is no upper limit to height. Some breeders are working to develop taller dogs in the Tsang-Khyi style. These dogs often have more developed flews and wrinkles as well.

All-black or black and tan dogs were the most common in the Himalayas, but all colors are seen today, including white, cream, gold, red, and even steel blue to silver. The coat is dense, forming a ruff around the neck and a shawl or mantle down the spine, and the plumed tail curls over the back. Tibetan Mastiffs are not recommended for hot, humid climates but do well in the weather patterns of their homeland, which vary from extreme cold to hot, dry summers.

CHARACTER/TEMPERAMENT

The Tibetan Mastiff must be socialized, and care must be taken that he does not protect his family's children too aggressively. He can be a tremendous barker if left outside at night, and he was bred to have a distinctive deep, bell-like bark. Young dogs can be aggressive chewers and diggers; in fact, chewing is the primary reason Tibetan Mastiffs are turned in to rescue efforts. They are not usually aggressive toward other dogs, and they have a low prey drive.

Tibetan Mastiffs are regarded by some as better property guards than livestock guard dogs. Traditionally these dogs were kept in the home or camp and let loose at night as sentinels and guardians. The best job for the Tibetan Mastiff is as a farm and home guardian rather than a remote livestock guardian. The American Tibetan Mastiff Association (ATMA) has published this statement on the use of the breed as a livestock guardian:

"ATMA does not recommend that the Tibetan Mastiff be used as a full-time livestock guardian dog. There is little evidence that the Tibetan Mastiff has ever been used as such in Tibet. The ATMA believes that there is too great a potential for danger with a Tibetan Mastiff raised without human interaction and socialization. Tibetan Mastiffs are a guardian breed and without proper socialization, may become unacceptably protective of their property for modern society. ATMA recommends that people interested in using a Tibetan Mastiff as a livestock guard dog, should plan on having the Tibetan Mastiff live with them in their house, and have access to the livestock. In addition, ATMA does not recommend that a Tibetan Mastiff ever be kept in an unfenced area."

SPECIAL CONSIDERATIONS

There are some health concerns in the breed. All breeding dogs should be screened for hip and elbow dysplasia. As is common in large dogs, osteochondrosis and panosteitis can be a problem as well. Tibetan Mastiff breeders also need to screen for hypothyroid and canine inherited demyelinative neuropathy (CIDN), a nervous system disorder that has been identified in the breed.

ON THE FARM

Torsten and Phil Sponenberg
Beechkeld Farm, Blacksburg, Virginia

Dr. Phillip Sponenberg is the technical director of the American Livestock Breeds Conservancy and a world-recognized expert on preservation genetics. It should come as no surprise, therefore, that he and his wife, Torsten, preserve three old strains of Tennessee Myotonic goats. They have about a hundred animals on their farm. After losing goats to coyotes, the Sponenbergs tried different breeds and crossbreeds of livestock guard dogs.

Although most of the dogs guarded well, none of them fit the farm's particular needs. One profusely haired breed presented problems both with incessant barking and with the need to be sheared annually. Another breed did not exhibit proper submissive behavior around stock. Some very young puppies the Sponenbergs brought in were already oversocialized to people and were never content to stay with their stock, even when placed with goats and older guardian dogs. Some dogs guarded only by default or were simply territorial rather than nurturing toward their charges. The crossbred dogs were unpredictable. Other dogs were overly aggressive and untrustworthy with people — even their owners!

Phil and Torsten wanted a specific dog. "We were interested in shortish hair, smallish size, and a reasonably nonaggressive dog," Phil says. They also wanted a purposeful guardian that would interact well with the goats. After some investigation, the Sponenbergs imported five Karakachans (Bulgarian Shepherd Dogs). More Bulgarian dogs have followed these initial immigrants to Virginia.

Phil reports, "[The Karakachans] offered a real choice to me that seemed otherwise unavailable — smaller size, smoother hair (somewhat long for my tastes, but they shed out on their own without much help from me), and safe for people, and, as an added bonus, they don't bark as much. These dogs start barking, then figure out if whatever they noticed needs further attention or not. So far, all of them have been real thinkers, which is counter to my experience with my previous breeds and the crosses. They also stay put! They have been ideal for our situation, though they would *not* be ideal for all situations. I think this is an important concept — one dog cannot fit all bills!"

The dogs at Beechkeld Farm generally work in pairs. Despite their uneven socialization backgrounds, the imports all proved safe around livestock. When he has puppies to train, Phil agrees that an older dog is the best mentor. "Some of them are stellar at this job," he observes. Phil also moves young dogs around, changing partners and fields, in order to make them more adaptable to future situations.

Besides coyotes, the dogs eliminate raccoons and skunks, which carry rabies. Phil believes this job is important, and it is another reason why he uses dogs instead of guardian donkeys or llamas. "The dogs are a further barrier between rabid wild animals, our herds, and us."

Phil believes the greatest challenge a new owner faces is "learning that the dogs generally do things their own way. I suspect that some folks just cannot get their dogs to work because they are not letting them do the job. Some people find them too frustratingly independent."

What is his advice to a potential owner? "See the parents of the pups and talk to the breeder as well as previous customers. Not all dogs work in all situations! I always prefer it if customers look around — then if they choose what I have, I feel it is an informed decision and is likely to work out."

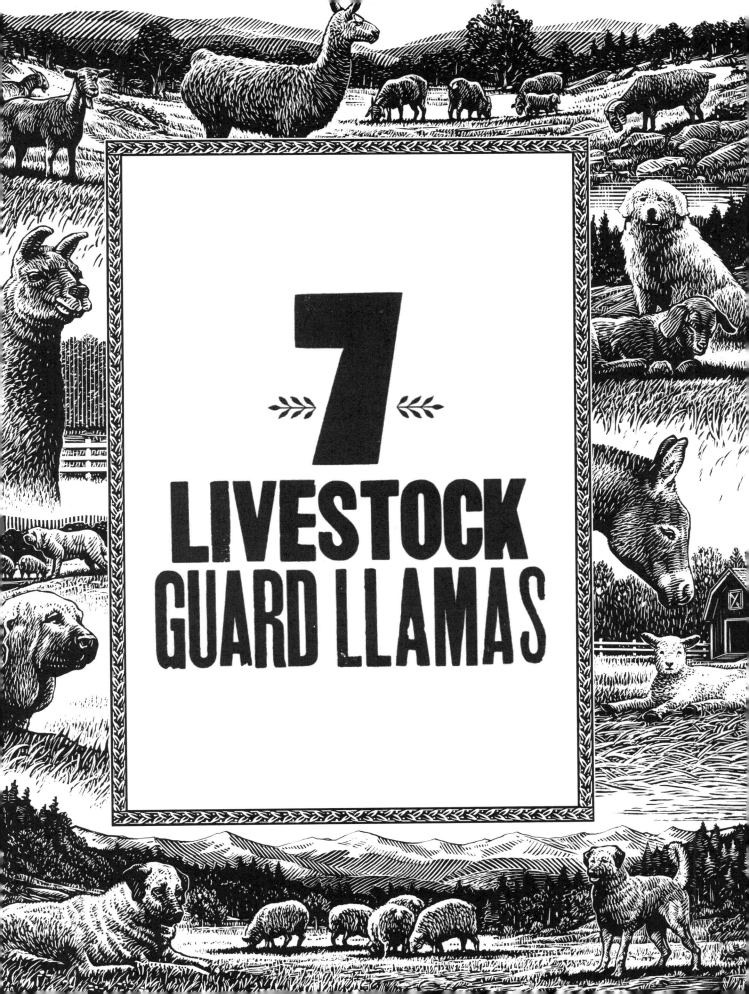

7
LIVESTOCK GUARD LLAMAS

Although their use as livestock guard animals is a fairly recent development, llamas are increasingly being used to protect flocks and herds of other species, grazing and living with them. Llamas are naturally social and when they are the only member of their species on pasture, they tend to associate with their pasture mates. Many llamas are both naturally and reliably aggressive toward coyotes and dogs as well as other predators, and their behavior extends protection to the other animals they live with.

There are several good reasons why a llama might be a better guardian for your stock and your situation than a livestock guardian dog. The primary reason cited by most users is the lower level of maintenance. A llama can eat or graze the same foodstuffs as the flock or herd. Llamas generally do not challenge fencing, so you don't have to prevent roaming, the primary cause of most livestock guardian dog losses. Obviously llamas don't bark, which may be a very important factor if you are worried about disturbing your neighbors. Llamas typically pose no potential threat to human beings, which may also be an important consideration in your decision-making process. Llamas bond fairly quickly to their stock; there is no long training period. Your stock, especially if they are skittish or wary of dogs, will probably accept a llama guardian more readily. Finally, some people are just not comfortable with large, powerful dogs.

Llamas cannot provide the same level of protection that livestock guard dogs do. For example, llamas can usually defend themselves against a single dog or coyote but are less able to defend themselves against groups of canines. But only you can make the decision about what type of guardian is best your situation. A llama may be a preferable choice for you, depending on your predator threats.

Origins and History

Llamas, alpacas, vicuñas, and guanacos all belong to the family Camelidae. Llamas are relative newcomers to agriculture in North America, but in a biological view, they were actually coming home when they were first imported here. In reality, all of the camelids originated in North America. The ancestors of the Bactrian and dromedary camels probably found their way into the Old World across the Bering Strait land bridge.

Those camelids that stayed behind and found their way to the mountains of South America adapted to life at 4,000-meter elevations through the development of small, elliptical red blood cells that possess a greater capacity for carrying oxygen. Llamas and alpacas were domesticated in the Peruvian Andes 6000 to 7000 years ago. Llamas became the pack animals of the high Andean plateau, but in this herding

Llamas bond fairly quickly to their stock without a long training period.

economy, they were also important sources of meat, fat, leather, and fiber. In fact, Andean people used all parts of the llama, including the bones and manure.

The exact ancestry and relationship of the members of this camelid family have been difficult to determine since their discovery by Western biologists. The most recent genetic analysis seems to show that the vicuña *(Vicugna vicugna)* and alpaca *(Vicugna pacos)* are most closely related. The llama *(Lama glama)* appears to have descended from the wild guanaco *(Lama guanicoe);* however, all of these camelids can successfully interbreed. It appears that before the arrival of the Spanish in 1528, there were various domestic breeds of double-coated llamas and single-coated alpacas, including the *chaku* llama with heavy fiber, the *kcara* or *ccara* llama with a finer fiber, the *huacaya* alpaca with fine crimped fiber, and the *suri* alpaca with long, fine fiber. Today these names are still used to describe the modern types or breeds. There may have been even more breeds or fiber types in the past.

The llama and alpaca populations before Spanish colonization in areas of modern-day Peru, Bolivia, Chile, Argentina, Ecuador, and Colombia were estimated in the tens of millions. Since the Spanish colonizers preferred their own domesticated animals, 90 percent of the llama and alpaca population was eliminated during the following century in favor of sheep, and wild vicuñas and guanacos were hunted into near extinction. One result of the diminished population was increased crossbreeding between llamas and alpacas and their various breeds or types. Crossbreeding with llamas continues today as the South Americans attempt to meet an increased market for alpacas and their fiber.

LLAMAS IN NORTH AMERICA

Small numbers of llamas, guanacos, and various crossbred camelids were imported to North American zoos beginning in the late 1800s. In the

Many alpaca breeders utilize guardian llamas to protect their valuable animals. Alpacas, while exhibiting many of the same behaviors as llamas, are too small to be effective against larger predators.

Llama Vocabulary	
CRIA	A baby
DAM	A female with a cria
GELDING	A castrated male
KUSH	To lie down
MAIDEN	A female that has never been bred
MATRON	A female that has given birth
OPEN FEMALE	A female that is not pregnant
SIRE, STUD	A male that is used for breeding

early 1900s William Randolph Hearst imported llamas to stock his San Simeon estate in California. Descendents of these llamas later made their way to the Catskill Game Farm and the Patterson llama herd. The Patterson ranch herd in Oregon served as the foundation of most llama breeding in this country, and its breeding practices and record keeping helped establish the International Llama Registry and most other efforts important to the llama industry. These original imports formed the basic population of llamas in North America until new imports arrived in the 1980s.

Alpacas were first imported into the United States in the 1980s and '90s, and their population exploded rapidly. There are now about 100,000 llamas and perhaps 60,000 alpacas in North America. It is interesting to note that in contrast, there are only about 60 guanacos and a handful of vicuñas in private ownership in the United States. The areas with the largest populations of llamas are California, Oregon, Washington, Colorado, Texas, Wisconsin, and Ohio. Ohio is also home to the largest number of alpacas.

Behavioral Characteristics of Llamas

In South America the wild camelids live in groups of five or six female adults guarded by a dominant male that is aggressive toward intruders. These groups live in distinct territories, marked with piles of dung. The climate is harsh and the forage limited, so it is important that the male drive away not only rival males and predators but also other grazing groups. The male is the primary aggressor in these situations, although females also engage in aggressive behaviors against threats.

The natural predators of the camelids include the mountain lion or puma and the variously named Andean, South American, red, or culpeo fox. The Andean fox, which can weigh up to 30 pounds, functions much like the coyote in the local ecosystem.

Llama Spit

Domesticated camelids generally do not spit at humans unless they are frightened. Warning signs are the ears being pinned back and the chin lifting. Though smelly and disgusting, there is nothing actually harmful about llama spit; it is just partially digested forage.

Condors prey on llama newborns, and the camelids behave defensively against them as well. Today free-roaming dogs are also major predators of wild camelids.

Camelids communicate with each other through body language and a variety of vocalizations, including a shrill alarm call and low humming. Camelids have acute vision and are curious and inquisitive about anything new or unusual. Camelids spit to show dominance, during social disagreements, or if they feel threatened. Females also spit to deter amorous males.

Camelids are highly social animals and do not like to be alone. Most camelid breeders will not sell a single animal unless it will be used as a livestock guardian and kept with other animals. Your sheep or goats are a llama's new herd mates; do not deprive him of their company. He may become quite agitated in their absence. A llama kept alone will become seriously depressed.

Curiously, llamas do not generally touch or groom each other. Since llamas have an attached tongue, they cannot extend their tongue out far enough to lick each other or anything else. Mother llamas do touch their babies by nudging them and sometimes gently sniffing them.

HOW GUARD LLAMAS WORK

Llamas are not used for predator control in South America. They were first used as guard animals in North America in the early 1980s, after shepherds noticed that predator losses were diminished when llamas were pastured with sheep. Sheep producers in Australia and New Zealand soon after began using llamas for protection against dingoes, feral dogs and dingo hybrids, and foxes. As producers reported their successes, various llama-use surveys and studies were initiated in the early 1990s. More comprehensive studies are under way at present, and from them we may learn more about how to choose proper animals as guardians. It is becoming evident that careful screening and testing of potential guard llamas increases their success rate.

Mature llamas are protective of their young, their herd mates, and their territory. Individual llamas may act in different ways toward threats, though most tend to be aggressive toward canines, including coyotes, foxes, and dogs. A llama's first response to a perceived threat is a high-pitched alarm call or scream, and posturing. If the threat persists, the llama will usually spit, then move toward the predator, and finally chase, paw at, or kick the predator. Charging or attacking the intruder is the most common method of defense. Some llamas will also place themselves between the predator and their companions or herd them away from the predator.

In 1990 researchers at Iowa State University surveyed users of guard llamas with sheep flocks to determine the llama's interaction with predators. The predators included dogs, coyotes, foxes, and bears. About one-third of the guard llamas indicated the presence of a predator by one or more "alert" behaviors: watching intently, giving an alarm call, or herding the sheep. Only 8 percent of the llamas placed themselves between the predator and the flock. However, nearly two-thirds of the llamas ran toward the predator and chased it. About 20 percent of the llamas actively kicked or pawed at the predator. Some of these owners reported that their guard llama had killed coyotes, muskrats, and woodchucks. Only 3 percent of the owners observed their llama walk or run away from a predator. Some owners have observed that female llamas are more likely to stay with their herd mates than to charge a predator, but this is not documented by research.

Some llamas assume leadership of their flock of sheep and become proactively territorial, patrolling fence lines and observing the area from higher ground, much like their wild relatives. The fact that their heads are about six feet up from the ground

Alpacas as Livestock Guardians?

Although alpacas are occasionally advertised for predator control, the vast majority of alpaca breeders in North America themselves utilize livestock guard dogs or llamas to protect their valuable animals from predators. Alpacas behave in the same territorial and protective ways as llamas but are quite a bit smaller, reaching 100 to 175 pounds in weight and about 36 inches in height at the withers.

Alpacas are being promoted as livestock guardians in New Zealand and Australia, primarily to guard newborn lambs and kids against fox predation. Early research has shown that the presence of alpacas reduces fox predation. Neither of these locales has predators larger than free-roaming dogs. Producers are using from one to six alpacas in their lambing pastures but removing them after weaning the lambs.

Many alpaca breeders are opposed to the idea of promoting the alpaca as a livestock guardian. Some believe that when confronted with a threat, alpacas tend to behave more like sheep than like llamas, fleeing from predators. Others maintain that dogs, usually in groups, are the primary predators of sheep and goats and that an alpaca stands no chance of successfully confronting dogs. In North America the combination of larger predators and high alpaca prices will probably prohibit the widespread use of alpacas as livestock guardians.

However, if your only predator threat is the fox and you would enjoy the alpaca fiber, an alpaca will probably provide some protection for your sheep or goats.

gives them good visual ability. Most importantly, most llamas are willing to accept other species as their herd companions. Studies have demonstrated that single guard lamas are more successful in preventing predation than multiples, since single animals will focus on the flock rather than each other. We take advantage of these behaviors when we use the llama as a livestock guardian.

One recent study by M. C. Cavalcanti and F. F. Knowlton found three traits that correlated with llama aggressiveness toward dogs: alertness, leadership, and weight. Alertness is observable as a still or frozen posture with the head held high and the ears erect and forward. Leadership includes not only having the herd follow the llama but also the protective behaviors described earlier. Finally, studies have shown that heavier llamas display higher levels of aggressiveness toward dogs, as compared to lighter llamas. Weight is often linked to age and maturity in llamas, but not always. Interestingly, llamas that were more aggressive in their normal behaviors toward humans were not more aggressive toward dogs.

Best Chance for Success

Llamas are believed to be most successful in pastures up to 300 acres in size; however, there are many variables to consider, including the roughness of the terrain, the vegetation, and the amount of predator pressure. Llamas appear to be less successful in dense vegetation or on open range. In larger or open-range areas llamas whose guarding behavior keeps them with their stock seem to be more successful than llamas that are more territorially watchful and remove themselves from the stock; however, many owners report that llamas that assume watchfulness from a distance are often successful as well.

Llamas have successfully guarded cows with calves, sheep, goats, deer, and poultry. Most producers who use llamas report coyotes as their primary predator worry. When William L. Franklin at Iowa State University surveyed 145 sheep producers who used guard llamas, 80 percent reported that their guard llamas were either effective or very effective at reducing coyote predation. More than half of the producers reported that their losses fell to zero. All

---

Pros and Cons of Livestock Guard Llamas

PROS
- Similar maintenance and feeding requirements to those of sheep and goats
- Relatively easy to fence in pastures
- Generally calm temperament
- Little threat to neighbors
- Does not roam, dig, bark, or chew on wood
- Less prone to accidental death than dogs
- Long working life
- Able to guard sheep, goat, cattle, deer, or poultry
- More easily accepted by dog-wary livestock
- Produces fiber
- Bonds to livestock quickly
- Fits predator-friendly guidelines

CONS
- Vulnerable to wolves, bears, mountain lions and packs of dogs or coyotes
- Does not provide protection against small predators such as raccoons or opossums or large birds
- May not adjust to living without other llamas
- May not accept herding or livestock guardian dogs working with the flock
- May injure or harass livestock
- May interfere with birthing process
- If not well trained or socialized, adult males can be dangerous to humans

of these producers continued to use other predator control methods in combination with their guard llamas. The producers reported an overall 5 percent mortality rate of stock with their guard llamas. This study also showed that predation was considerably less with a single llama than with two or more.

Llamas themselves are vulnerable to predation by dogs, coyotes, and other large predators. Even a single vicious dog can severely injure or kill a llama. The most dangerous dogs to llamas are the bull-baiting breeds that grab the nose and suffocate their prey. Although some owners swear that their llamas have deterred bears and coyote packs, the odds are not good for the llama in these situations. Llamas truly have no defense against groups of dogs or coyotes, wolves, bears, and the big cats. Some breeders report that llamas instinctively flee mountain lions, which are their natural predators. And the reality is that most llama and alpaca breeders themselves use livestock guard dogs to protect their valuable animals.

Some llama breeders will place a llama in a guardian job only if the predators are no more serious than foxes or the occasional solitary coyote. Some will recommend that you come to the llama's assistance whenever he gives an alert. Be aware that some llamas will not give a loud alarm call to alert you that they need your help with a predator, unlike livestock guard dogs that will bark loudly. Sometimes a llama can work with a guard dog; the llama can then serve as an excellent sentry but can flee to safety with the sheep or goats while the dog confronts the predators.

Choosing a Guard Llama

Expect to pay from $500 to $1,500 or more for a neutered male llama. Females usually cost more since they have reproductive ability. Either may make a suitable guardian, although gelded males are most commonly used because they are larger and less expensive. Male llamas weigh 300 pounds or more and stand 40 to 44 inches tall at the withers and 5½ to 6 feet at the head; females are slightly smaller. A female llama can be more nurturing than a gelding, especially if she has been used for breeding, although some owners believe that a female is less likely to chase a predator. A retired breeding female can be a very attentive guardian and may be available at a reasonable price. To be a good livestock guardian, a llama must be a mature animal not younger than 18 to 24 months. Llamas have a life span of 20 to 25 years and you can expect a long working life from your llama.

Just like everything else in life, you get what you pay for. In this case, you should look for an adult llama, which means the breeder has invested at least two years in feeding and caring for him. Buying an adult llama allows you to evaluate his behavior more accurately than you could with a younger llama. An older llama also has learned to accept regular handling, shearing, toenail trimming, and vetting. In many cases an inexpensive llama will have behavioral or other problems, while a more expensive llama from a reputable breeder will pay for itself in many ways.

Experienced breeders have a wealth of knowledge, and they will be able to help you learn to care for your llama or head off small problems before they become big ones. Llamas are occasionally offered for sale at livestock auctions; however, at an auction you will not be able to assess the llama's behavior around humans or livestock and its potential aggressiveness toward canines. Llama rescue groups report that very few of the available rescue llamas are suitable for work as guardians, but it is worth exploring this option if you are interested. There will be an adoption fee.

If you already have a predator problem, you should definitely look for an experienced guardian llama, which will probably cost more. On the other hand, if you are not experiencing serious predator problems at the present but would like the reassurance of a guard animal, you could choose a slightly younger or inexperienced animal and allow him to grow into his role.

Plan to make several visits to see the variety of llamas available to you and to gain some experience in handling them. If you know someone familiar with llamas, take him or her with you to help you pick a healthy animal with good conformation. Look for a clean, well-run farm and a breeder who is experienced with llamas. Ask for references or contacts from others who have purchased a livestock guard llama from the breeder. A good place to find a reputable breeder is through the various regional or national llama associations and their Web sites. Be honest about your plan to use this llama as a livestock guardian. Some breeders specialize in raising llamas for this purpose, and they can help you select an appropriate guardian. You may even be able to purchase an experienced livestock guardian or a llama that has been housed with the type of livestock you own.

A reputable breeder will give you a signed sales contract or bill of sale, a health record, and a guarantee. Try to purchase a bale of the hay the llama has been eating, so that you can make his transition to your feed slowly and safely. Make sure you agree upon any delivery terms. Fortunately, llamas can travel in horse, livestock, or utility trailers, covered pickup trucks, and even vans. (A trained llama will conveniently fold up or kush.) If the animal is coming from a distance, you may be able to contract with a nationwide animal hauler.

If you have any questions about the health or soundness of a llama that you might purchase, you can pay a veterinarian for a prepurchase health check. Arrange to be present for the exam if you can, so that you can observe the llama being handled by a stranger. Before you purchase a llama, confirm that your own veterinarian will treat llamas or that you can locate a veterinarian who cares for llamas. Before you buy, ask the breeder for the animal's routine worming medications and schedule and a record of its immunization history.

WHAT TO LOOK FOR IN A LLAMA

It is extremely important that you do not buy a very young llama or one that was raised apart from his own species. Llamas must spend the first year and a half to two years in the company of other llamas so that they develop both mature territorial instincts and appropriate llama behavior. Llamas should not be weaned until six months of age. Avoid bottle-fed llamas or llamas that received too much human attention when young, as they may have bonded with people rather than other llamas. Overindulgence or lack of limit setting by humans may result in a mature llama that views humans as competitors and exhibits inappropriate and possibly dangerous behavior.

If you buy a male, he must be neutered at least 90 days before purchase and he should probably have his fighting teeth removed. Do not buy a gelded male if he was ever used for breeding or was only recently gelded. Do not buy an intact male under any circumstances. Intact males fight with their heads, necks, teeth, and chests, and may challenge your authority. Furthermore, intact males will probably attempt to breed your livestock, causing serious injury or death to sheep or goats. Even some gelded males have been known to attempt to breed livestock. The management of intact males is best left to llama breeders.

Although most llamas have pleasant personalities, you should be very cautious when choosing a guard llama, especially if this is the first llama you have owned or handled. Do not buy a llama that screams or spits at all humans, paces his fence line, or does not allow people to enter his pen. Do not buy a llama that bothers your feet, tries to chest-butt

Llamas are highly social animals and do not like to be alone. They are protective of their herd mates and their territory.

you, or plants himself in front of you, forcing you to go around him. Llamas are curious and will often sniff you, but be extremely cautious of the llama that pushes his face at you or follows you around demanding attention. This is not friendliness but potential aggression. Avoid a llama that is overly protective of his food or challenges someone cleaning up his manure pile. Most llamas will accept a few pats on the neck or back. Some are more "cuddly" or "grouchy" than others, but the best guardians are those that are independent or even aloof toward humans. Under no circumstances should you buy a llama that makes you nervous or uncomfortable, especially if you have never owned or handled a llama before.

Look for a llama that is halter trained, accepts brushing, and is accustomed to general handling. Ask the breeder to let you catch any llama you are interested in purchasing; be cautious of buying a llama that is haltered before your visit. Practice leading and grooming the llama with the breeder's assistance. Have the breeder demonstrate toenail trimming, if possible, or make plans to have assistance the first time you need to do this task yourself.

Don't be alarmed if the breeder uses a chute or a catch pen to restrain the llama for trimming, but be aware that you may need to do the same at your home. It should go without saying that if llamas are handled only during toenail trimming, shearing, vaccinations, and worming, they will not have a positive attitude toward people.

If you are fortunate to find a llama that is already with stock, observe his behavior. Is he relatively easy to contain with fencing? Does he graze with his stock without herding them unnecessarily? It is normal to observe the llama leading, following, or walking and grazing in the midst of his flock mates. He may also use a higher area of the pasture as a lookout point, separating himself from the flock. All of these activities indicate an appropriate relationship. Do not buy a llama that paces his fence lines looking for other llamas rather than staying with or guarding his stock.

Here are some questions to consider:
- Has the llama has been present when sheep or goats have given birth?
- Does the llama appear alert and curious about disturbances?

- Does the llama show aggression toward dogs or give an alarm call? If you're not sure of the answer, ask the breeder if you can bring a strange dog into the llama's line of sight or pasture to evaluate his response.

Single llamas are generally recommended over multiple llamas so that the llama will bond with his charges rather than his fellow llamas. Early studies proved that a single llama guards better than multiple llamas. Where single llamas are kept in separate pastures with a common fence, the llamas are observed to spend considerable time socializing with each other across the fence.

In cases of owners who have successfully used more than one llama, they report that one llama will herd the stock away from the predator while the other llama will confront it. Alternatively, the llamas will take turns grazing and standing guard or patrolling the pasture boundaries. Using two females seems more common in these cases. Though some people have used two gelded males successfully, two males, even gelded, will tend to fight if there are differences in age or size. If you are purchasing two llamas from the same breeder, you can make sure they are compatible before purchase.

If you own farm or herding dogs, find out whether your llama will accept them before you purchase it. If the breeder has dogs, ask whether the llama accepts those familiar dogs. Otherwise, be very cautious when using your dogs in the llama's enclosure. Remember, you want your llama to be aggressive toward dogs and coyotes. Some llamas will come to accept familiar livestock guardian or herding dogs in their pastures, while others will need to be removed when you are working with dogs.

TYPES OF LLAMAS

While llamas can be divided into different types, they are not strictly speaking classified as distinct breeds. There is some evidence that in the past these types may have existed as landrace breeds in South America, but the widespread destruction of the native llamas and alpacas during the colonial era resulted in extensive crossbreeding. With the increasing interest in breeding for specific types in North America, these types may develop into standardized breeds in the future. At present, both llamas and alpacas are classified by fiber type rather than appearance.

A llama's type is not related to his potential success as a livestock guardian, but coat information

Many llamas develop close and even playful relationship with their charges. They seem to enjoy the company of lambs and kids.

is included here so that you understand the care needed for various types of llama fiber. Some llama coats require very little care, which might be attractive to you. Alternatively, you may be interested in using the fiber from your llama. If so, you have some interesting choices in types of fleece.

Ccara

Until the 1980s the majority of the llamas imported into North America and Europe were traditional working animals, primarily from Peru. Today about half of the llama population in Peru is still of that type. Because this sparsely coated llama provided the original image of this exotic creature, individuals of this type are often called classic llamas, or Ccara, in North America and Europe.

The Ccara has short, flat hair on his head, legs, and sometimes even the neck. He has a double coat with outer guard hair and underwool that sheds each year. The guard hair provides protection from rain and snow. Some Ccara coats are very short, with a large amount of guard hair and very little underwool. Because of its guard hair content, Ccara fiber was traditionally used for rougher products such as ropes, rugs, and heavy carrying bags.

The Ccara coat can usually be maintained with an occasional grooming to remove dead undercoat and any mats that might form. The shortest coats are really wash and wear. It is not necessary to shear a true Ccara coat unless it has been totally ignored and is impossible to comb through. When a Ccara is shorn you can really see that his coat is very sparse, especially in comparison to the dense coat of a true woolly llama. If a Ccara must be shorn, he will need protection from the sun, inclement weather, and biting insects until he regrows some coat.

Ccaras were selected primarily for size and weight-carrying ability, so they tend to be larger than the heavily wooled types. In addition, due to better nutrition and other factors, Ccaras in North America tend to be taller than in their native land. In South America Ccaras average 38 inches at the withers, while Ccaras in North America and Europe can stand up to 44 inches tall at the withers and 5½ to 6 feet tall at the head.

Curaca

The Curaca is regarded as a subtype of the classic or Ccara llama. The Curaca is a medium-wooled animal whose head and lower legs are covered with short, close hair rather than wool. It may not be possible to comb out the coat of a Curaca llama, and he will require shearing. Since the composition of the fleece is so variable, a Curaca may have lesser or greater amounts of guard hair.

Tampuli

The Tampuli, as it is called in southern Peru and Bolivia, is the true woolly llama. Bolivia is home to almost 70 percent of the world llama population, and about 80 percent of Bolivian llamas have this medium-length to long wool. About 50% of Peruvian llamas are the Tampuli type as well. The fleece of a woolly llama is dense and does not shed every year. Although you will see woolly llamas in full coat for the show ring, it is difficult to maintain this coat, so it is definitely recommended that the owner of a woolly guard llama shear him yearly. Tampuli llamas can be further divided into two sub-types: Tapada and Lanuda.

Tapada llamas have a small amount of wool on the head and on the leg below the knee on the leg, with a heavy coat on the body. The woolly coat can grow to a length of 7 to 14 inches.

Lanuda llamas have a great deal of wool on the head, often in fringes on the ears and head, and abundant wool down to the ankle. Lanuda fleeces show such little difference between the guard hairs and the undercoat that the guard hair need not be removed when the wool is used for spinning.

Suri

There is some evidence that a now-extinct variety of llama with extremely fine fiber once existed in precolonial South America. This llama may have been an ancestor of the Suri llama, or perhaps the Suri resulted from crossbreeding between llamas and alpacas. In any case, the Suri, with his size and striking coat, is an impressive animal. His distinctive fleece falls from a midline part on the neck through to the tail. The long locks of fine fiber have no crimp or loft. These locks may be straight, wavy, twisted, or curled. Breeders keep their show animal in a long, full coat. The Suri must be sheared.

The Suri is generally considered impractical for work as a guard animal, since his fleece provides less protection against the weather and requires considerable care to keep it in good condition.

ON THE FARM

Cheryl Lavooi
Woolly Manor, Republic, Washington

Cheryl Lavooi lives in the mountains of northeastern Washington State, where she raises Shetland and Icelandic sheep and teaches spinning, fiber processing, weaving, and other crafts. Cheryl faces predation from coyotes and loose domestic dogs, as well as mountain lions and bears. For the last three years she has used a very large llama named Bandit to protect her flock. She chose a llama specifically because she did not want a huge dog as a guardian. Also, because her business is home based, she is available to monitor the stock during the day. Bandit vocally alerts her if he detects anything unusual. Cheryl further ensures her flock's safety by keeping them in areas situated right next to her home at night and for winter pasturing, her times of greatest predator threat.

Cheryl observed one encounter when a dog got into a pen with the sheep. "Bandit gave his call to send the sheep to the barn and *ran* with head down to chase the dog! That dog belongs to a neighbor, and he has never tried to go into the pen again, but Bandit still watches him carefully whenever he sees him."

Bandit towers over both Cheryl and her small Shetland sheep. As a first-time llama owner Cheryl was glad that her new guardian was comfortable with being handled, behaved appropriately toward people, and was trained to lead. She likes taking Bandit along when she takes her sheep out to graze on unfenced pastures. In fact, Cheryl believes she is able to do this without a herding dog because the sheep stay close to Bandit. At times he has called an alert and herded the sheep back to the barn. He will also stake out safely on a long rope while the sheep graze nearby.

Cheryl's greatest challenge with Bandit has been trying to trim his feet and shear him herself. She has decided that the best approach for her is to have the professional shearer do both tasks. Cheryl has also learned an invaluable tip for catching a llama: She uses a stake line or long rope with one end fastened to the fence to form a triangle across a corner of a pen or paddock. Keeping the rope taut, Cheryl walks slowly around Bandit, holding the rope about chest high, until she is standing at his head. She laughs, "Works every time — even when he's feeling frisky!"

Crossbred Llamas

Crossbred llamas can have any type of fleece and coat length, but their fleece tends not to shed and is difficult or impossible to comb. A crossbred llama will need to be sheared every year in hot climates, although he may be able to go two years between shearings in cooler areas depending on his rate of growth.

Introducing Your Llama to Livestock

If your livestock animals have never seen a llama before, they may be fearful of or unsettled by their new guardian for some time. Your new llama will also be nervous or uncertain. Initially the llama may be happy to see some companions or simply curious and might run toward his new herd mates, while the sheep or goats may run away. Even if he does not run toward the flock, your llama probably will be interested in them. Some llamas, however, are neutral toward other species and may be quite calm when introduced to them. Your llama will most likely become more composed and settle in within a few hours, although he may need about a week to become fully adjusted to his new home.

The common advice has been to place your llama out in the field with his new herd mates, and he will bond with them in a few hours. Research has shown, however, that you will have more success if you first spend a few days acclimating the llama and the stock to each other in a small area. If your pasture or range area is large, your chances of long-term success will increase if you pen the llama together with your animals for several days in a smaller corral. Although some users claim the llama will guard the sheep better if he is introduced to them when lambs are present, research hasn't supported this. Owners do report that llamas, especially females, are very interested in lambs.

Llamas and Horses

Be very cautious when introducing llamas to horses that are unfamiliar with the species. Frightened horses may run and injure themselves. Experienced owners recommend that you introduce horses and llamas to each other very slowly over a month or so. If you have horses on your property or if you plan to pasture them near llamas, please exercise similar caution.

If your stock are extremely nervous, pen your new llama next to the animals and entice them to feed near each other to encourage interaction. Flighty sheep breeds or goats may not only be more difficult for your llama to guard but also may take longer to become accustomed to their guardian. At first your llama may be more comfortable with humans and seek out your attention. He needs to bond with his new herd mates rather than you, though, so you need to refrain from giving him too much attention until this bonding has occurred.

Basic Training for Your Llama

Most new llama owners will not have had any experience in caring for or leading one of these large animals. Acquiring some of this experience at the place where you purchase your llama or from someone familiar with llamas is really beneficial. Knowledge of specific techniques or suggestions, as well as a measure of confidence, will increase your chances of success.

USING A CATCH PEN

Catching a llama in a large open field is a difficult task, even for several people. If your llama lives in a very large pasture, you will find it easier to catch him if he is accustomed to entering a catch pen in his area. Llama breeders utilize such pens to handle their animals. A catch pen can be 12 to 20 square feet in size. If you use the catch pen to feed the llama, he'll be happy to enter it. It will also be a good place to secure him if you need to separate him from his flock or herd while you or others work with the stock.

Llamas that are uncomfortable or unaccustomed to being caught should be approached with care. In their panic they may attempt to jump or go through a fence or otherwise injure themselves. If possible, catch him inside a shelter with solid walls. Move slowly. Once you have him cornered, stop to let the llama calm down before you begin handling him.

PROPER USE OF HALTERS

Do not use sheep or foal halters on a llama. Instead, use one of three types of halters designed specifically for llamas: fixed noseband, X-style, and two-way adjustable. With a fixed-noseband halter the crown piece is adjustable. Fixed-noseband halters come in different sizes and are inexpensive, but they are the least comfortable and safe for the llama. X-style halters are not adjustable, since the crownpiece and nosepiece form one continuous loop, but they are comfortable and usually stay in the proper position once fitted correctly. Two-way halters are adjustable in both the nose and the crown pieces. They not only are comfortable when adjusted correctly but also are safe and effective when used in llama training.

A llama halter must fit properly or it is dangerous and frightening for the animal. A well-fitted halter always stays up high on the nose bone, very close to the eye. It has enough slack in the nose piece that the llama can open his mouth and chew naturally but not so much that it slips over the nostrils. Llamas cannot breathe solely through their mouths; if the halter covers their nostrils, they can suffocate. The crownpiece controls how far the noseband slips down the nose, not the nosepiece.

Some halters are not well proportioned and will be impossible to fit correctly and safely on your llama. A poorly fitting halter may actually cause your llama to misbehave. Check that halters do not become too tight during the year due to the growth of fleece. Tight halters can cause abscesses or other skin problems. Never leave a halter on a growing llama.

HALTER TRAINING

If your llama is not used to being haltered or has had previous bad experiences with haltering, you may need to train him to accept the halter. First, he needs to be restrained in a catch pen or chute. Open the noseband very wide. It will be easier to tighten it later than to try to loosen it once it is on the llama.

- Hold the noseband above the llama's grain, so that he eats by placing his nose through the halter. Allow him to try this once or twice.
- Then, without the grain, drape a safe catch rope high around his neck to help guide his nose into the halter.
- Stand on his left side, close to his front. Hold the halter by the check piece and allow the crown piece to hang straight down toward the ground.
- Slip the nosepiece high up on the llama's nose, while holding the buckle in your right hand.

A correctly fitted halter sits up high on the nose bone, very close to the eye.

- With your left hand, reach under his chin and bring the crown piece up over the top of his neck and into your right hand. Do not reach around the back of his neck, as this makes many llamas nervous.

You should remove the llama's halter in the catch pen rather than the pasture, so he does not come to associate running away with the halter removal. You should make this experience calm as well. Unbuckle the halter, but hold the two pieces together in your right hand. With your left hand, ease the nosepiece off. You can still steady the llama with the closed halter and then gently release him.

Because llamas are so curious, it is not advisable to leave halters on unsupervised llamas on pasture. In addition, do not leave llamas tied to a fixed object, unless they have been trained to accept tying. As with sheep or goats, being tied is easier for a llama to learn when he is young and small, but a calm approach over time should be successful if you need to be able to tie him. If a llama must be tied, an elastic extension on the tie rope may prevent a broken neck or other injuries.

LEADING YOUR LLAMA

Use a long line with a knot in the end for leading your llama. Attach your lead rope to the side ring of the halter. Hold the lead line with two hands, but keep your left hand down near the end of the lead. You can hold the extra length of the lead in flat loops in your left hand as well. If you pull on your llama's lead rope, he will probably pull back. It is more effective to turn his nose to make him take a step. Stay in front of your llama as you lead him.

If your llama bolts, you will have a greater advantage if you are farther away from him. Then take a few steps closer to him so that you aren't pulling. Rough or hard pulling will only make the llama more fearful and may hurt him. If you lose the line, wait until he stops running before you pick it back up.

RESTRAINT TECHNIQUES

A well-trained llama can be restrained with a halter and lead rope. If necessary, you can stand your llama next to a solid wall and distract him by tightly grasping the base of his ear. It is not necessary to twist the ear, but you do need to hold on firmly. You can also try having someone hold his tail.

You can build a chute if your llama requires restraint during routine procedures. You can lead the llama into the chute and then back him out when you're done. A chute can be built conveniently into the catch pen or elsewhere. The sides should be about 5½ feet tall and the chute should be about 2 feet wide. Pipe gates, lumber, or plywood can be used for the sides. A stall or horse trailer can also be used as a chute.

However you restrain him, note that when a llama kicks he usually does so like a cow, kicking his hind leg forward and outward. Only rarely will he kick backward.

A veterinarian can also administer a sedative or other anesthetic to your llama if he resists routine tasks like nail trimming or shearing.

If your llama requires restraint for common procedures, you can build a chute from pipe gates or lumber.

8
TAKING CARE OF YOUR LLAMA

Although llamas are similar to cattle, sheep, and goats, there are important differences between them. Llamas do regurgitate and chew their cud, but they possess three stomach compartments instead of four. Instead of hooves, llamas have padded feet with large toenails that must be trimmed regularly. And their major arteries are not always in the same locations as in other stock.

Llamas also have some important sexual differences. Female llamas are induced ovulators, which means they will not come into heat without the presence of a male. They can, however, be bred year-round. Males are not completely mature until they are two to three years old, when the connective tissue between the prepuce (the sac holding the penis) and the penis disappears, allowing the penis to become fully erect. However, some young males are still capable of penetration, so do not assume that it is safe to house an intact male with females at any age. Llamas are generally gelded at eighteen months to two years of age, since earlier gelding may cause abnormal skeletal development. Waiting too long before gelding is not advisable, since the young male will be exposed to mature hormones.

WHAT TO FEED YOUR LLAMA

Llamas can do well on arid pastures with an average protein content of 2 to 4 percent. In fact, lush pastures or high-protein diets may cause them to become overweight or ill. A llama can eat a variety of hay or pasture plants, although alfalfa is considered too rich and some vegetation is poisonous

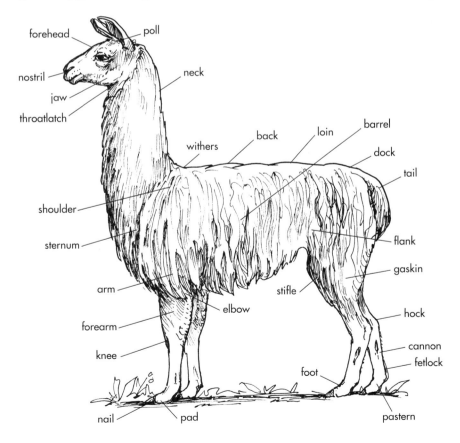

to llamas. You can check your state department of agriculture for specifics in your area. When not on pasture, a llama will need approximately one bale of hay every week to ten days, or about four pounds daily. He may be fed as little as once a day if he is not grazing. He will also do quite well with the large round hay bales used to feed stock.

Hay or pasture, water, and access to loose minerals and salt are all a nonbreeding llama requires. Llamas cannot survive on snow and do require water daily, unlike their camel cousins. Llamas cannot lick hard salt or mineral blocks; give them instead a loose salt or mineral mix or soft mineral or salt blocks. Llamas also require copper in their mineral mix, which may be toxic to sheep, so you need to provide access to a llama mineral mix in a sheep-proof feeder above the reach of the flock. Check with your state agriculture authorities to find out which mineral formulations are appropriate with your pasture or local hay.

Supplements or grains are not necessary except for pregnant, nursing, or working pack animals. Llamas can bloat if they eat very large amounts of grain. If your llama will have access to livestock grain or supplements, be aware that he can choke on straight pellets. He can also bloat if he drinks too much water after consuming pellets. A mixed-grain feed, with or without pellets, is preferable. Competition for food encourages llamas to eat too rapidly, also putting them at a risk for choke. The best solution is to provide multiple feeding areas or stations if you feed grain or to create a barrier to prevent the llama from entering sheep or goat grain areas.

Llamas can be quite fussy about wet hay, and they will waste less hay if they eat from flat-bottomed feed troughs rather than hayracks. The flat-bottomed troughs will also eliminate the risk of llamas catching and trapping their heads in the narrow slats of some feed racks.

If your stock is crowding the llama away from food, causing aggression or weight loss, place his food in a feeder higher than the sheep or goats can reach. This will remove a major source of your llama's stress. If your llama is with horses or cattle, you can use a catch pen to feed him, placing a pole or poles across the opening low enough to prevent the stock from entering. A llama can crawl under an amazingly low barrier if motivated.

PROVIDING SHELTER

Obviously the type of shelter your llama needs is dependent on your climate. In some areas llamas can live comfortably outside all year so long as they have natural or artificial shade and protection from rain. If you need to build shelter to protect your llama from either hot or cold weather, he will prefer an open area with large doors or windows. He will avoid small, dark sheds or stalls. You may need to construct something specifically for your llama if you use low buildings or other shelters for your sheep or goats. Shelters for inclement weather should face away from prevailing winds, while shelters for hot weather should be oriented to catch breezes. Shade shelters, in fact, can be open on all sides.

It may be necessary to install a feeding or mineral mix station for your llama that is out of reach of your stock. A plastic horse stall feeder works well for this purpose.

Your llama will also appreciate a dry area to eat and kush, especially if your ground is wet or muddy for long periods of time. Dry areas will also help prevent foot problems. Llamas can develop fungus or "rain rot" skin problems if they are kept in overly wet conditions. Straw bedding is not necessary if his area is dry. If you intend to use his fleece, note that wood chips or sawdust will badly contaminate it. Sand will also collect in the fleece, but it can be removed.

If the temperature frequently drops below 15°F, your llama will need an enclosed shelter or access to the barn where his stock is housed. Use deep straw bedding for added warmth, although he will also appreciate the body heat of his flock or herd mates.

FENCING REQUIREMENTS

Llamas do not require specialized fencing. Though llamas are capable of jumping a 4½-foot fence, a contented llama can be contained with a 4-foot fence. Since llamas are territorial, a single llama in a livestock guardian situation does not generally challenge his fencing. However, llamas can crawl under fences if they are determined to escape, so the bottom barrier needs to be no higher than 12 inches from the ground. Llama breeders most often use 5- to 6-foot-tall fences made of appropriately sized field fence or no-climb fence.

The major concern with llamas and fencing is their propensity for sticking their head and necks through things. For this reason, fences made of wire strands can be a problem. In addition, the llama's fleece may prevent him from being shocked by electric fence. Some owners place small pieces of masking tape or plastic ribbon on electric fence wire to encourage llamas to touch it with their nose and learn that it shocks.

Nonelectric high-tensile wire fences can be especially dangerous to llamas, as they can catch their heads and long necks between two or more wires. Llamas can also stick their head, necks, and limbs through livestock panels or wire mesh fencing if the holes are large enough.

Barbed wire is very dangerous for llamas, especially to their protruding eyes. As with other types of wire fencing, llamas will stick their heads and necks through barbed wire, which can cut them badly. In addition, llamas with long fleeces can become entangled in barbed wire, since they are fond of rubbing against fencing.

Because they are so curious, llamas will stick their heads through dangling loops of wire, rope, or string, so be sure your fence doesn't have any. Cover your gates with wire mesh to prevent similar stuck-head problems. Llamas will also chew on cedar posts and rails.

Routine Health Care

You may need a catch pen or chute to restrain your llama for routine care, which includes trimming toenails, dental care, grooming, and shearing if necessary. Llamas also must be wormed regularly, and they require yearly vaccinations. Consult your veterinarian for specific recommendations for your area.

VACCINATIONS

No vaccines have been formally approved for llamas, but the common stock vaccines are being used regularly. Regular yearly vaccinations include

> ### Blood Work
>
> **The location of** the jugular vein is different in llamas than in other livestock, and it can be difficult to locate the major arteries. If you're not experienced working with llamas, have someone knowledgeable about llamas perform blood collections and give intravenous injections.

Clostridium perfringens type C and D, tetanus toxoid, and killed rabies, if necessary in your area. Your veterinarian may recommend other vaccinations.

The most suitable areas for subcutaneous shots are in front of the hind leg, in front of the shoulder, and behind the front leg. Intramuscular shots can be given on the upper hind leg or above the elbow.

EXTERNAL PARASITES

Several external parasites may possibly trouble llamas, including flies, nasal bots, and sarcoptic mange mites. Llamas should be checked and treated for whatever external parasites are present in your area, although they are not overly susceptible to lice, fleas, or ticks.

Medicated shampoos and livestock dusts do not harm the llama's fleece. Pour-on formulas do damage fleece, however, and have not been proven safe for llamas; most are actually dangerous, often causing paralysis or coordination problems. Most fly preparations formulated for horses can be safely used on llamas.

INTERNAL PARASITES

Llamas are susceptible to coccidia, liver flukes *(Fasciola)*, meningeal worms *(Parelaphostrongylus)*, necked strongyles *(Nematodirus)*, nodular worms *(Oesphagostomum)*, stomach worms *(Haemonchus, Ostertagia,* and *Trichostrongylus)*, tapeworms, threadworms, and whipworms *(Trichuris)*. Consult your veterinarian for specific treatments for these parasites. Deworming is recommended at least twice a year; quarterly is preferable.

As with vaccines, no dewormers have been approved for llamas, but several products appear to be safe. Llamas share some species of these parasites with other livestock and wildlife. In particular, white-tailed deer are hosts for meningeal worms, which in llamas may cause paralysis. If you live in an area with meningeal worm, Ivomec injectable will combat it but must be given every four to six weeks. For other deworming needs, ivermectin paste, moxidectin gel, and fenbendazole paste are safe and will combat most of the parasites that trouble llamas. You should alternate dewormers to decrease the possibility of parasites developing resistance to them.

Do not use Ivermectin Plus, pour-on dewormers, or other dewormers unless recommended by your veterinarian. Pyrantel pamoate is not effective in ruminants. Do not use drench guns or syringes designed for other livestock. If you use paste or suspension dewormers, insert them into the cheek pouch of the llama's mouth, not the center of the mouth or on the tongue, where the llama can spit them back out. Some llamas will eat pastes or liquid dewormers when they are mixed with grain.

TRIMMING FEET

Your llama will probably need his toenails trimmed to prevent lameness. Some llamas need to be trimmed only yearly, while others need trimming every other month. Some owners have their llama's toenails trimmed when he is sheared. Do not attempt to trim your llama's toenails without a lesson from his breeder or a trained person. With instruction and a well-trained llama, you should be able to perform this routine task, but it takes practice.

You can maintain your llama's acceptance of toenail trimming with a few considerations. Be patient,

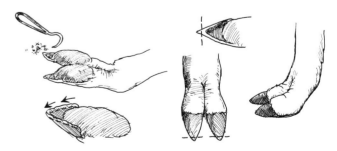

Scrape dirt out of the crevices of the foot, trim the sides of the toenail level with the pad, and trim and round off the tip of the nail.

calm, and relaxed when handling your llama. Remember that llamas are very uncomfortable with having their feet handled, probably because they fight each other by biting at each other's legs. And it is easier to trim the nails when they are moist.

Some people like pruning-type nippers for trimming llama nails, while others prefer farrier-style horse nippers. Attend to other grooming tasks before attempting to pick up his feet. You may need to brush his legs several times before he will let you pick them up. When you pick up a foot, do not pull the foot out to the side, since this causes discomfort. Trim off small pieces rather than risk cutting into the quick of the nail.

You can also attempt to trim his nails without picking up his feet. Restrain him in a chute, if possible, on a hard, flat area. You should be able to trim away small pieces of the nails by keeping the nippers parallel to the ground. If you are patient, you can do a decent job of trimming in this manner.

However, if your llama is very uncooperative about having his nails trimmed, or if he has overgrown and neglected toenails, it is best to call for a professional or have your veterinarian sedate your llama for the procedure.

CARE OF TEETH

Llamas are modified or pseudo-ruminants with a hard upper pad or gum instead of upper front teeth. With their flexible upper lip and lower front teeth, they grab forage and tear or rip it off into their mouth. They have both upper and lower molars for chewing forage. The "baby teeth" fall out and are replaced by permanent teeth usually by the time llamas are five years old.

Male llamas also develop six to eight large, sharp canines (four on the top and two or four on the bottom) variously called fighting teeth, wolf teeth, or fangs. These teeth begin to erupt at about age two and are present by age four. Females have

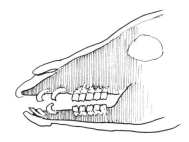

The sharp canine or fighting teeth of a male llama may need to be removed or blunted.

smaller canine teeth that usually pose less danger. Many owners have their veterinarians remove the fighting teeth to prevent aggression injuries. Occasionally the fighting teeth grow back and may need to be trimmed or blunted periodically until they stop growing at seven or eight years of age. If your gelding or female is nonaggressive, these teeth may not be a problem.

Older llamas, like all other ruminants and horses, may lose or wear down their incisors and molars. If the teeth develop sharp edges (called points) or uneven wear, they may cause a llama pain or make eating difficult. A veterinarian can float (file) points, as is done for horses. As a llama ages, his incisors may grow too long or uneven, requiring cutting down or rasping to allow him to eat more efficiently or comfortably. Eventually an older llama may lose so many molars that he will need a special diet to maintain a healthy weight. Signs of such dental problems include difficulty in chewing, excess salivation, or weight loss in a llama of any age.

Llamas can develop abscesses, especially in the lower molars. Symptoms include avoidance of cold water or grain and dropping quids (wads of partially chewed forage) out of the mouth. There may also be swelling along the lower jaw or face. After diagnosis, a veterinarian may be able to treat the llama with antibiotics if the abscess is small. Surgery or extraction may be required for larger abscesses.

GROOMING AND SHEARING

Llamas do not particularly enjoy grooming, as it is not a natural activity among them as a species. The classic llama coat can be maintained with a thorough grooming twice a year. A slicker brush will take care of dried mud, dead fiber, and vegetation or debris. A comb or rake will remove the underwool. Comb lightly through the fiber, pulling outward, not down. If the guard hair is somewhat tangled you can use a spray-on conditioner or detangler fro horses. Llamas have very sensitive skin. Never comb a llama whose coat is so matted or tangled that it will not comb easily. Llama breeders sometimes use blowers to remove dirt and sand from short coats.

Llamas are susceptible to heat stress and do not shed their dense undercoats, so most animals with medium-length to long fleece should be sheared in the spring in hot, humid climates. An unshorn coat is not only a major contributor to dangerous heat stress but is also unhealthy. The matted undercoat and tangled guard hairs trap dirt and moisture. Large mats can pull painfully at the skin and cause serious irritation. Feces may build up around the tail and rear legs, attracting flies and other parasites.

Although there are different shearing cuts for llamas such as the blanket, barrel, show, or pack cuts, the most practical cut for the guard llama is a complete shearing of the entire fleece, including the neck, legs, and tail if necessary. Leave a coat of ½ to 2 inches to protect the animal from sunburn and provide some insulation against heat or cold.

If you raise sheep, you can ask your shearer whether he or she also handles llamas; some do and some don't. You may need to arrange for someone experienced in shearing llamas and alpacas. In extremely hot climates you may need to shear your lama twice a year or when the coat has grown to about 2 inches in length.

Specific Health Concerns

Llamas tend to be healthy animals. Normal rectal temperatures vary from 99 to 101.8°F, although higher temperatures can be normal during hot weather. Normal respiration is 10 to 30 breaths per minute and a normal heart rate is 60 to 90 beats per minute. Llamas should eat every day. Watch for llamas that refuse to eat, although llamas do not always lose their appetites when they are not feeling well.

Llamas are noted for behaving stoically even when ill or in pain. If you pay attention to your llama's normal behavior, you will be able to notice something out of the ordinary when he is ill. Signs of an ill llama include lying on his side, failure to rise from a kush, lameness, excessive rolling, frequently getting up and down, or standing with his head down and his back hunched. Although llama pellets may change color with different diets, diarrhea is abnormal. Urine may be clear to yellow in color, but white or chalky sediment is abnormal.

Using Llama Fiber

Llama fiber comes in a large variety of colors. It has less lanolin (or grease) than sheep wool and is light in weight. While true sheep wool is solid, llama wool is technically a type of hair because it is hollow. Llamas other than the Ccara type can grow about 4 inches of fleece per year. The soft fine underwool does have some crimp or waviness but less elasticity than sheep wool. Many spinners mix a small amount of sheep wool into their llama fiber for spinning yarn.

Llama fleeces change with age in both males and females. The best fleeces come from younger animals. Fleeces that have matted or felted or contain burrs and weed seeds are less valuable to spinners. Unless your llama has an unusual fleece, it is probably no more valuable than a sheep fleece.

CHOKE

Llamas can clog their airways by eating too rapidly, which they generally do when they feel a sense of competition from other animals. Llamas can choke on anything from supplements to hay, although straight pellets are the most frequent cause since they swell with moisture. If choking is not resolved, asphyxiation is a real possibility. Symptoms include coughing and gagging. In an emergency you may be able to massage a choking lump loose, although it is recommended that you have a veterinarian pass a tube down the throat to clear the obstruction. It is also possible that food can be aspirated into the lungs.

Llamas are less likely to choke on pellets if they are mixed with a coarse grain. If you must feed straight pellets, spread them out in a large, flat feeding trough or place a number of large, smooth rocks in the feeder to slow your llama down. Since guardian llamas do not require grain or pellets for good nutrition, it may be preferable to feed them to your livestock in an area your llama cannot enter.

COLIC

Colic is a general term for discomfort or impaction in the gut. Causes can include a change in diet, too much grain, or a heavy parasite load.

Untreated colic can be fatal. Symptoms include the llama lying flat on his side, changing sides restlessly, kicking or nosing at his abdomen, groaning, making repeated attempts or straining to pass manure, failing to eat or chew cud, and grinding the teeth. A veterinarian can use mineral oil, by stomach tube or enema, as well as medications to treat colic.

HEAT STRESS

The llama's natural environment is the high, arid Andes. He will have difficulty in extreme heat and/or humidity, especially over an extended period of time. Overweight, pregnant, old, ill, or upset llamas

Llamas can behave stoically when ill. A seriously ill llama may stand with his head down and his back hunched.

are even more vulnerable. Heat stress is a serious health issue for llamas. Nevertheless, llamas have been kept successfully in all areas of the United States, and there are precautions you can take to avoid problems with heat stress.

- Shearing the llama, especially the ventral abdominal area, before the summer is helpful; the abdominal area can be resheared before especially hot weather. Long, tangled, or matted coats prevent the llama from dispersing heat. Leave just 1 to 3 inches of fiber.

- Make sure to provide shade for the llama with trees, shade cloth, or a roofed shelter. Be aware that your llama is much taller than your sheep or goats, so he requires several more feet of vertical air space than they do. Try to locate shade in areas that catch the breeze. If necessary, you can use fans and misters to provide preventive relief. Spread sand several inches deep under shade trees and periodically dampen it.

- Llamas will make use of natural and artificial water sources, even a child's plastic wading pool, so be sure to provide plenty of fresh, cool water for drinking. You can also add electrolyte mixtures to his water.

- Be cautious when transporting llamas during very hot or humid weather. Try to avoid moving a llama from a cooler climate to a hotter one in the summer.

Pay attention to your llama's normal behaviors so that you can detect the small changes that reveal the onset of heat stress. The earliest symptoms include a lack of interest in eating and drinking and depression or apathy. You may observe your llama lying near his water trough or defecating while lying down. Watch for signs of colic. Panting, a drooping lower lip, salivation, trembling, weakness, and abnormal breathing are more serious symptoms, as are a rectal temperature greater than 104°F and a respiration rate greater than forty breaths per minute. Increasingly serious symptoms include edema and the inability to stand.

If you observe early signs of heat stress, you must cool your llama down by hosing him with cool water or placing him in water so that he is drenched down to the skin. Just wetting the fleece may cause it to mat, creating further insulation. Keep him in the shade, and use a fan to circulate air around him, or take him into an air-conditioned room. More extreme treatments include alcohol rubs, ice-water enemas, and intravenous fluids. Your veterinarian can also prescribe anti-inflammatory drugs (such as Banamine), injectable B or E vitamins, selenium, probiotics, and antibiotics.

HYPOTHERMIA

Hypothermia in llamas is possible during periods of extended cold. Shivering is the first visible sign, followed by labored breathing. Normal rectal temperature should not be below 90°F. Llamas are sometimes reluctant to drink very cold water, which may result in dehydration.

If your llama is shivering and possibly dehydrated, offer him warm water immediately. Bring him into a warm, sheltered area, or use heat lamps to bring his temperature up. If he drinks and his shivering stops within a half hour, you may not need to consult a veterinarian.

Solving Problem Situations

If you have a llama, you may encounter one or more of the situations or problems described below.

PROBLEM: Aggression toward Humans

SOLUTION: A llama might display aggressive behavior toward humans if he is removed from his peers at too early an age or if he is not taught to behave properly around people. Llamas need to be handled appropriately to learn good manners and maintain acceptable behavior. Do not let your llama push you or bother your feet with his head or neck. Do not reward attention-seeking behaviors. Use the words *stay back* to enforce your personal space. Always move deliberately and calmly around your llama. Just as with handling horses, dogs, or other animals, if you allow bad behavior it will persist and become more entrenched.

Five percent of owners report that their llama became so protective that it was difficult to work with the livestock without removing the llama from the flock. Most owners take the precaution of removing the llama when routine livestock care may upset him.

For more help in training or handling your llama, consider a training book or seminar. For truly serious behavior problems, you need an experienced trainer or handler. The rare llama with berserk or aberrant male syndrome was most probably raised or handled improperly. This animal is a danger to humans since he may bite, knock people down and attack or wrestle with people's legs.

PROBLEM: Aggression toward Stock

SOLUTION: Aggression toward and attempts to breed the livestock are the most common problems reported by those who have attempted to use a guard llama. The scent of an ewe in heat is similar to that of the female llama, making sheep especially vulnerable to this problem, although male llamas have attempted to breed other species as well. Male llamas breed by mounting prone females, and sheep can be injured or killed by this behavior. In the Iowa State University study, these problems occurred in 25 percent of the intact male llamas and 5 percent of the gelded llamas.

To prevent this problem, buy a guard llama that has been proven safe with sheep. Do not buy a female llama to keep with a gelding with aggressive tendencies in an effort to distract him away from your sheep. He may continue to bother your sheep, and he will certainly injure the female llama through repeated attempts at breeding, since she will never become pregnant.

PROBLEM: Aggression toward Working Dogs

SOLUTION: Although a guard llama can become tolerant of herding dogs used with his stock, it is common for him to chase the herding dog if lambs are present and they bleat in fear. If you expect your guard llama to accept your herding dog, realize that you are asking him to see a difference between one dog chasing his stock and strange dogs chasing his stock. This may be possible with some individual llamas, but not with others. It would be better to remove your llama before using herding dogs on stock, especially when stock will be subject to medical tasks or shearing. If this is not possible, you should expect your guard llama to be aggressive toward the herding dog, at least initially.

Although llamas will generally become used to family dogs that behave calmly around their stock, it is advisable not to allow them in the llama pasture. If you expect your llama to be an effective guard, he should not be habituated to tolerating dogs with his stock. It is quite true that dogs often guard llamas themselves, but in that case the llamas are relying on the dogs for protection. Sheep producers who have used guard llamas report that the llamas have sometimes attacked and injured family dogs. Some llamas will never become used to the family dog and will chase or attack it at every opportunity.

PROBLEM: Birthing Season Issues

SOLUTION: Use caution during the first birthing season of any species, since the llama may attempt to herd the newborn away from its mother. Be aware that the llama may actually be trying to help the newborn find its mother. Many llama owners report that llamas are very interested in lambs and other newborns. Lambs are often observed playing with the guard llama.

PROBLEM: Failure to Work

SOLUTION: Some llamas will simply not prove to be good guardians. Either they will not stay with the stock or they fail to exhibit any guardian traits. They may either ignore dogs and other predators or run from them. They may not exhibit any territorial instincts. These problems are reported more often with young llamas than older ones. If a young llama does not improve as he becomes older, it may be necessary to find him a new home, so it is important to ask if your breeder will exchange or take back a llama that does not work as a guardian. The best prevention for this problem is to buy an older, proven llama as a guardian.

Llama Fertilizer!

Llamas tend to create and use communal manure piles for their pellets, which are similar to those of deer, sheep, and goats. If your llama does return to use the same toilet area every day, the pellets are easy to collect. This manure is an excellent odorless fertilizer that, because it is low in nitrogen, can be applied directly in gardens without first being composted.

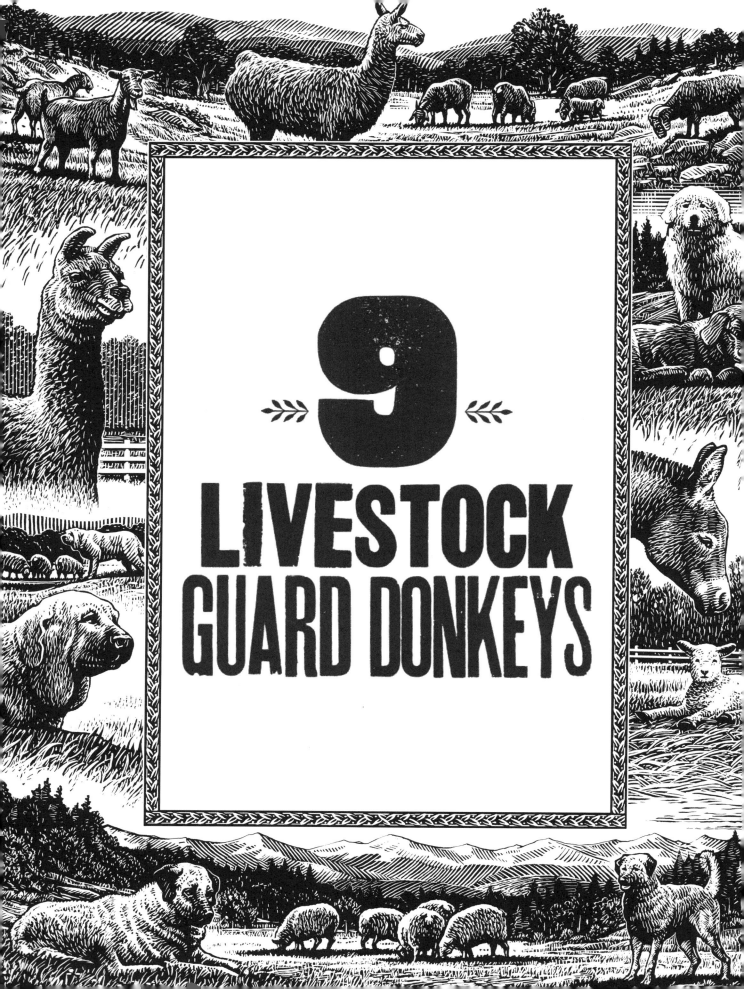

9
LIVESTOCK GUARD DONKEYS

Single donkeys can be used as livestock guardians, grazing with, living with, and protecting the herd or flock of another species. When they are the only member of their species in the group, they tend to associate with their pasture mates. Many donkeys are both naturally and reliably aggressive toward canines, and their behavior often extends to protecting the other animals they live with.

The use of donkeys as livestock guardians is fairly recent in North America, but these animals were being used to protect livestock from large carnivores in Namibia more than a century ago. This practice has been revived with some success as part of conservation initiatives in Africa. Donkeys have also recently seen use as livestock guardians in Switzerland, being used in place of dogs, which can be seen as threatening and frightening to passersby, in areas open to day hikers.

Origins and History

There were once several races of the African ass *(Equus asinus)*, which is a separate species from the horse. Today's domesticated ass, known as the donkey, is primarily the descendant of the Nubian wild ass *(Equus asinus africanus)*, which became extinct in the 1960s. Other African wild asses still exist, though they are endangered. The Egyptians and other peoples domesticated the donkey as long as 5,000 to 6,000 years ago. The donkey is, in fact, the only large domesticated animal to originate from Africa. The lowly donkey is still the common working equine of the Middle East and Africa.

All of the wild ass species have a large head, long ears, a short, upright mane without a forelock, a wispy tail, slender legs, and small feet. The wild ass appears in shades of gray or dun with a dark dorsal stripe down the back and across the shoulders forming a cross. At times there is faint striping visible on the legs. There are lighter-colored areas on the muzzle, around the eye, and on the underparts. Although domesticated donkeys once looked like wild asses, today they come in many other colors and patterns. The most numerous colors are slate gray, brown, and black, but sorrel, bay, ivory, roan, and spotted are also seen. The hair of the coat can be flat, curly, long, shaggy, smooth, or wiry. Donkeys have a very long life span of 30 to 40 years.

Donkey Vocabulary

BURRO	A Spanish term for the small working donkey. Today this word refers to the ferral, free-roaming burro population descended from Spanish stock.
DONKEY	This term first appeared in England and may have derived from a combination of *dun,* meaning "grayish brown," and *kin* (later *ky*) or "little." *Donkey* is preferred over the term *ass* in North America.
HAND	A unit of equine measurement equaling 4 inches
HINNY	A hybrid offspring of a stallion and a jennet
JACK	A male donkey
JACK COLT	A male donkey under three years of age
JENNET, JENNY	A female donkey
JENNY OR JENNET FILLY	A female donkey under three years of age
MULE	A hybrid offspring of a jack and a female horse

After domesticated donkeys spread across Europe, six of them traveled to the New World on the second voyage of Columbus. Donkeys and mules were soon bred in large numbers in Mexico for use in the Spanish New World empire. These hardy little beasts of burden were soon found throughout the West, where some found their freedom and easily adapted to the climate and geography. Burros, as they were known in the colloquial Spanish, established large feral populations.

By the 1920s feral burros and horses were being rounded up for dog food. Feral burros came under the protection of the U.S. Bureau of Land Management after the Wild Free-Roaming Horses and Burros Act was passed in 1971. In the 1970s the burro population was estimated at about fourteen thousand. Today that number has been reduced to less than five thousand, and the Bureau of Land Management eventually plans to maintain the wild burro population at 2,750.

Donkeys and mules have been important partners to humans in both agriculture and transportation in North America. In 1920 there were 5.4 million working mules on the continent, before automobiles and tractors replaced them. Even by 1930 there were still some 48,000 donkeys engaged in agriculture. Since the 1950s the domestic donkey population has remained somewhat stable at about 55,000. What has changed is the exploding population of the Miniature Mediterranean donkey, which now makes up half of the North American donkey population.

Worldwide, the vast majority of the 44 million donkeys are hardworking beasts of burden, but in the United States today most donkeys are pets or are used for recreational riding, driving, packing, and showing. Since the 1980s, for the same reasons producers turned to livestock guard dogs, interest grew in using donkeys as livestock guardians.

Behavioral Characteristics

The wild North African asses were strongly territorial in nature, much more so than wild horses. This behavior is believed to be a response to the harsher environments they occupied. The jacks defended large areas, especially the valuable water supplies. Males and females tended to come together only during breeding season, with adult animals living alone or in small groups during the rest of the year. These groups were highly variable and often changed their members from day to day. Jacks fought to establish dominance over each other through biting, kicking, and striking.

Similarly, feral burros in the Southwest do not live in well-organized social groups. Jennies and foals stay together for up to two years, but otherwise the social groupings are very loose. Burros will graze together for a few hours or days but then go their own way. Due to their inherited social behaviors, donkeys are more likely than horses to be content living apart from other donkeys, but they do need some companionship.

Unlike ruminants, equines cannot retreat to a place of safety to chew their cud. Equines need to

Horse Socialization

Wild horses have social organizations very different from those of donkeys. Stallions protect stable groups of mares, foals, and juveniles. The horses within these bands form complex and intense social bonds with each other. These bonds may endure for years. Wild and feral horses are rarely territorial, except in unusual environments. Instead, most stallions protect a zone of intolerance around the band. This allows bands to come together to form large herds while still maintaining the integrity of the individual bands.

A jenny is naturally extremely protective of her foal. This instinctive behavior may extend to her herd mates.

spend long hours grazing, making them more vulnerable to predators. As a defense, speed and stamina became essential components of the equine survival system. Fleeing is the first and primary equine response to a threat, with teeth and hooves coming into play only when equines are unable to escape the predator. Experts believe the donkey's fight response may be triggered more quickly than the horse's because donkeys lived alone or in groups or two or three, and flight is not as successful as an escape technique for solitary animals as it is for a large herd. The natural predators of both wild and feral donkeys include canines and the big cats.

Other equine defenses include their senses of sight, hearing, and smell. Equine hearing is superior to humans' in both the lower and upper ranges. In fact, an equine that appears to be looking at a disturbance may actually be listening intently to identify it. Donkeys, with those huge ears, are well known for their excellent hearing. Equines also have an excellent sense of smell. Stallions can detect a mare in heat at a great distance. They identify each other, in part, by smell, which is why they will nuzzle to smell those they know.

Equines have both binocular and monocular vision. Monocular vision is the ability to see separately from each eye. This is an advantage for a grazing animal. Equines move their head and neck from side to side as they graze in order to enlarge their field of vision. However, they cannot see directly behind or in front of themselves, which is why experienced handlers know not to approach an equine from these directions without speaking to it. When something attracts the attention of a donkey, it will turn its head to focus on the object. This change from monocular to binocular vision causes the object to appear to jump. These vision characteristics lead, in part, to equines spooking or startling easily. Generally donkeys are not as spooky as horses and will tend to freeze rather than flee at a disturbance. Donkeys have larger eyes than horses, which provides them with a wider field of vision.

Most domestic donkeys are docile and friendly by nature. Donkey fanciers refer to their famed stubbornness as a "strong sense of self-preservation." Donkey owners report that you can't force a donkey to do something; she has to want to do it herself. Motivation is best provided by utilizing the donkey's own desires for food or social interaction. On the other hand, donkeys quickly learn to avoid things they find frightening or painful.

HOW GUARD DONKEYS WORK

Although the deliberate use of donkeys as livestock guards is relatively recent in North America, it is increasing rapidly. There is surprisingly little research into why donkeys work well as guardians and how to identify a good potential guardian. At present the number of producers who use donkeys as livestock guardians is smaller than those using either dogs or llamas, and donkey users are still reporting them less successful at preventing predation than either dogs or llamas. However, the lack of knowledge about choosing and training donkeys as guardians, as well as unrealistic expectations for donkeys to instantly bond with stock or guard stock over very large or rough terrain, has no doubt contributed to these mixed results. Increasingly careful selection and training of livestock guardian donkeys will improve this situation. We need to learn how to foster the donkey's inherent territoriality while acknowledging her less developed herd orientation.

Murray T. Walton and C. Andy Field have done the most comprehensive survey research on the use of guard donkeys. Their 1989 surveys of sheep and goat raisers who used donkey guardians in Texas revealed some interesting observations. The primary predators involved were coyotes, foxes, and dogs. Users appreciated the relatively low cost of donkeys, the easy maintenance requirements, and the fact that they could continue to use other predator control methods such as livestock protection collars, trapping, shooting, snaring, and denning. Only about 40 percent of owners rated their guard donkeys as excellent to fair, though there was a tremendous range in the situations the donkeys were expected to work in: from 100 to 1,000 acres and up to 1,200 sheep and goats. The average number of stock was 213, and most pastures fell between 200 and 600 acres in size. The lower numbers would seem to be a much more realistic match to the donkey's abilities.

Most of the users surveyed were using either a single jenny or a gelding. A few had a jenny and a foal. Those who had tried an intact male reported problems with overly aggressive behavior toward stock. Most owners had purchased their donkeys from private sources rather than through livestock auctions or the Bureau of Land Management adopt-a-burro program. Owners who were unhappy with

Donkeys are an alert grazing species, and individual animals will stay with their stock if they have become bonded to them.

their guard donkey complained that either the donkey did not guard the stock or it was aggressive toward the stock. Others reported that although the donkey attempted to drive off the predator, it was unsuccessful.

Donkeys tend to be an especially alert grazing species, and they are willing to interact with other livestock. Although a guard donkey often comes to accept the flock or herd as companions, it cannot be proven that the donkey is primarily and deliberately protecting the livestock. It is more likely she is protecting her grazing area and herself, though some would argue that jennies are maternal protectors and may extend this protection to their herd mates. Donkeys do not patrol the pasture. If a donkey is somewhat bonded to the stock, she will choose to stay with them as they graze, and the simple presence of a large animal among smaller sheep or goats provides some deterrence to dogs and coyotes. Many owners report that the sheep see the donkey as a protector and gather near or behind her if they are alarmed.

On the Attack

Most donkeys appear to have an instinctive dislike of canines, but there is no guarantee that a particular donkey will be aggressive toward dogs and coyotes. Some donkeys ignore intruders unless they themselves are threatened. Other donkeys will flee a predator. However, if they do attack an intruder, donkeys can be fearsome. They may charge the predator, using their teeth and all four hooves to attack. They will often attempt to bite the neck, chest, buttocks, and upper legs of the animal. They will use their front hooves to slash the predator's back and sides. They may also turn and kick at the animal's head. They will also chase the animal they are attacking, and they will be unaware of you or any other animal in their path. You cannot stop a donkey that is attacking, and you may be injured if you attempt it. Allow her time to calm down and return to her normal self before you try to approach or handle her.

Despite this aggressive reaction, a donkey will not be able to deal successfully with multiple dogs or coyotes, wolves, bears, or mountain lions.

At present there is only anecdotal information about how to identify a potential guardian donkey. Frustratingly, it appears that if the donkey is good, it is very good, and if it is not good, it is either useless or dangerous to the stock. In fact, users report everything from the complete elimination of predation losses to no change in predation rates to stock losses due to the behavior of the donkey. In the case of negative results, however, it often appears that users were too unrealistic in their expectations of the donkey.

CHOOSING A GUARD DONKEY

Donkeys are relatively inexpensive and easy to locate, but if possible, seek out those few breeders who are raising donkeys specifically for the purpose of livestock protection. They are making selections based on good guardian traits, the donkeys will have been exposed to the species they may protect since a young age, and the donkeys will be have been handled and gentled.

Choose a standard-size or larger donkey to protect your stock. If you have cattle, choose a donkey that is at least a little taller than your stock, such as the large standard. Do not use miniature donkeys to protect stock. They are undeniably cute, appealing, and easier to handle, but it is not fair to expect such a small animal to protect itself or your flock against predators. Dogs have mauled many miniature donkeys.

If you already have a predator problem, choose a donkey that is at least three years of age, past the playfulness of youth. Some users believe that a weanling (a donkey three to six months of age) should be taken from its mother and placed immediately with stock. This may be a good way to bond a donkey with stock, but this baby will not be an effective guardian for quite some time. In fact, she is vulnerable to predators herself. Since she is small and uncertain, she is also vulnerable to being bullied by her pasture mates. Also, young donkeys may play too roughly with lambs or goats. Donkeys often play by grabbing each other by the neck. This inappropriate behavior should be discouraged, since it could be fatal to smaller animals as the donkey grows in weight and size.

CHOOSING A JENNY OR JACK

A gelded jack might be your ideal purchase. A gelding typically has a more stable temperament than a jenny. In addition, a gelding is usually cheaper than a jenny. Intact jacks are more difficult to handle and are much more likely to be aggressive toward stock. They tend to be more headstrong, and they definitely bray more than geldings or jennies. An intact jack belongs with an experienced handler. If you are going to use a gelding, it should have been castrated at least ninety days before being placed with stock. A veterinarian familiar with donkey anatomy should do the castration.

A jenny also makes an excellent guardian. Some jennies are more unsettled or aggressive toward lambs or kids during their heat or estrus periods. Some users recommend a bred or pregnant jenny as a livestock guardian since she will be very protective and alert to potential predators. Her foal will be born with stock and can later be sold as a livestock guardian itself. Although a jenny with a foal will be extremely wary of predators and protective of her foal, she may also be aggressive toward stock if they are inquisitive about her baby. She may also ignore the stock in favor of her offspring. She will also find it difficult to actively guard the stock when she is heavy with foal.

Livestock Guard Donkeys

PROS
- Inexpensive
- Extremely long-lived, with a life span of thirty years or more
- Can guard sheep, goats, and calves against canines and possibly bobcats
- Similar maintenance and feeding requirements as pastured animals
- Relatively easy to fence in pastures
- Generally calm temperament
- Little threat to neighbors
- Does not roam, dig, or bark
- Less prone to accidental death than dogs
- Bonds to stock within a few weeks
- Compatible with other means of predator control
- Fits predator-friendly guidelines

CONS
- May not exhibit guardian behavior
- Vulnerable to packs of dogs or coyotes, wolves, bears, and mountain lions
- Ineffective against feral hogs
- Does not provide protection against small predators or large birds
- May not accept herding or livestock guardian dogs working with the flock
- May injure or harass livestock
- May interfere with birthing or breeding process

If you are interested in starting with a foal, you may be able to arrange to borrow or buy a pregnant jennet. The foal will grow up with the stock and will become closely attached to them. You can remove the jennet at weaning time (four to six months) and allow the foal to grow up with its stock companions.

WHAT TO LOOK FOR IN A DONKEY

Choose a healthy and sound animal. Her coat should be shiny, or at least free from sores and patchy hair, which may indicate problems. Her eyes should be bright and alert, and her ears should be forward or moving around, not continuously laid back flat against her head, which may indicate a sour temperament or pain. Her feet should be trimmed and she should appear friendly, indicating that she has been handled. She should be free of obvious conformation problems. Do not buy a lame animal, as she may never be sound.

If you are not familiar with equines, take a knowledgeable friend along to help you examine the donkey. You might choose to pay for a veterinarian's health check. You should receive a record of the animal's health care and a purchase agreement. If the donkey is being sold as a livestock guardian, you should receive a guarantee in the event that she does not work out well. Remember to locate veterinarian services for your new donkey before you need them in an emergency.

Choose a donkey that is friendly toward people and that you can handle. When you meet any new donkey, stand at least a foot in front of her, and place your hand out with your palm flat and facing up. Let the donkey smell you before you pat her. Do not pat the front of her face, since she cannot see your hand.

Be wary if the owner already has the donkey caught and haltered when you arrive. Ask that the donkey be released into a small area and see if you can catch and halter it safely. Try picking up her foot and cleaning it with a hoof pick. If you don't know how to do this basic task, ask the owner to teach you. If the donkey is not halter-trained, used to being tied and groomed, and used to a farrier trimming her feet, you will need to accustom her to these things. Be realistic about your abilities and time for this project. You may wish to purchase an already trained donkey.

Donkeys are large, strong animals, and if you have not handled them before, it is understandable that you may be uncertain around them. However, they are also sensitive, and nervousness in the handler will make the donkey more unsettled as well. Be honest about your abilities and experience. You may be able to arrange some handling and care sessions from your breeder or another knowledgeable individual.

Do not purchase an unmanageable donkey. You may receive advice that an untamed donkey will more aggressive toward predators. Please remember that you will need to work among your stock and move them without interference from an aggressive donkey. The donkey also needs regular care, and you will need to handle her to properly care for her. Allowing a donkey to live without regular hoof and medical care is unacceptable and inhumane.

Donkeys are individuals with unique personalities. They can be pushy or spoiled, timid or nervous, and sluggish or sullen. The ideal donkey will be friendly and trusting. Regardless of her personal traits, you should buy a donkey that you feel comfortable handling. Longtime donkey breeders believe that most donkeys can become good guardians except those that are extremely aggressive or extremely shy.

It may be possible to test the donkey's reaction to predators by placing her in a small corral or pasture and introducing a dog. Since some donkeys are violently aggressive toward dogs, take care not to endanger the test dog. Although the donkey may not alert well to a familiar dog, you should be wary if the seller claims that familiarity is the reason the donkey is not reacting to a particular dog. The donkey may instead not react to any dog.

You will need to arrange to transport your new donkey home. The seller may be willing to do this for a fee. It is possible to arrange for commercial delivery by a horse hauler, if necessary. Many donkeys are not well trained to load easily. It may be necessary to seek a calm, experienced person to help you load your new guardian.

You will need an interstate health certificate if you cross state lines. Most states require that donkeys being transported through them have a negative Coggins test, which tests for equine infectious anemia. Since this requires a blood test, either make arrangements to have the test done in advance of your transport date or make sure the donkey has a valid negative Coggins test already.

Cost of a Donkey

Expect to pay no more than a few hundred dollars for a gelded donkey, unless he is an experienced livestock guardian. Cheaper donkeys are available at auction barns or through the Bureau of Land Management wild horse and burro adoption program (see page 197), but in both cases you have no guarantee that the donkey will safely guard your stock. Buying from an established breeder means

ON THE FARM

Mary and Larry Limpus
Limpus Farms, Amsterdam, Missouri

After years of little predator threat, coyotes and roaming dogs have caused a steep increase in the problems faced by Mary and Larry Limpus. The couple raises Shetland sheep on 50 acres of rolling, rocky hillsides on the Kansas-Missouri border. Their working farm dogs, Australian Cattle Dogs and Australian Shepherds, were not able to deal with the problem. Although they have lost sheep in a pasture next to the house, their greatest losses have occurred in a pasture separated from their farm site by a railroad line. They decided to try two different options — donkeys and a Great Pyrenees — in different pastures.

Before the arrival of the donkeys, the Limpuses were losing a sheep every four to six weeks, and the carcass would be completely eaten. However, since the donkeys have been guarding the sheep, Mary relates that there has been only one loss, and that carcass was not fully eaten. She assumes the donkey ran the predator off during or after the kill. Mary and Larry believe that good fencing is assisting the guardian animals in doing their jobs.

The greatest problem Mary and Larry faced with the donkeys was getting the sheep used to their new guardians. As Mary says, "The donkeys liked the sheep better than the sheep liked them!" The solution was to create a small corral for a donkey in the sheep pen while they all became acquainted with each other. They have also been cautious about placing a new donkey with ewes and lambs.

In their experience, the Great Pyrenees did better with this job, while the donkeys were out with rams or other adults in a more remote pasture. Their Pyrenees is careful not to go into any pasture with the donkey, although their donkeys seem to tolerate the dog in other fields or on the farmstead.

you should get some sort of guarantee or right of replacement. In addition, your breeder may specialize in selecting stock with the appropriate traits for a successful guardian. The breeder also will probably be available to help you with donkey care and handling basics. You can find a breeder through the Web sites of the American Donkey and Mule Society or the Canadian Horse and Mule Association.

DONKEY BREEDS AND TYPES

Donkeys differ from horses in their conformation. Besides the obvious ears, the neck of the donkey is straighter than that of the horse, and a donkey does not have noticeable withers. The back is straighter, ending in a croup and rump without the muscling found on a horse. Donkeys lack a forelock (with the exception of the Poitou), and their mane is stiff and coarse. Many owners clip or shave the mane short. The tailbone is covered in hair, with a switch of hair at the end. Many donkeys have thinner legs than horses, although good bone structure is preferred. The donkey hoof is smaller and rounder in shape than a horse's hoof. And then, of course, there is that loud, raspy voice!

Donkeys come in a variety of sizes but only a few breeds worldwide. Most donkeys in North America are described by their size alone. The majority of the donkey population is a mixture of several landrace breeds from different countries that were brought to the New World in 1700s and 1800s. There are very few imported or purebred donkey breeds in North America.

Standard Donkey

The Standard donkey, which stands between 36 and 48 inches at the withers and weighs 400 to 500 pounds, represents the majority of the donkey population worldwide. In North America the Standard was created from a variety of imports over the centuries. Large Standard donkeys often contain some Mammoth Jack in their background; jennies measure 48 to 54 inches, while jacks can be as tall at 56 inches, and they weigh up to 800 to 900 pounds. The Standard and the Large Standard are the types most commonly used as guardians.

Mammoth Jack

The Mammoth Jack (also called the American Standard) is an American original created from several European breeds. The development of quality mules was accomplished in North America by importing and crossbreeding large Spanish and other European donkeys expressly for breeding excellent working mules. The result was the Mammoth Jack,

Donkeys vary widely in size. A Standard donkey stands 36 to 48 inches at the withers and a Large Standard between 48 and 56 inches, while a Mammoth Jack measures 14 to 16 hands.

the finest mule-breeding ass in the world, selected toward refinement, size, and conformation.

Mammoth Jacks are horse-size at 14 to 16 hands or more and 1,100 to 1,300 pounds. The Mammoth Jack is well balanced with a straight top line, a long, well-muscled croup, well-sloped shoulders, a wide chest, and well-sprung ribs. The legs are large and well formed. The head has a straight or slightly Roman profile. The ears can measure 34 inches from tip to tip. Black coloring with light-colored points was favored in the past, but light or reddish sorrel with a white mane and tail is popular for crossing on Belgian mares. Spotted and gray Mammoths are also available.

The Mammoth Jack remains the largest donkey in the world today. The population of Mammoth Jacks is estimated three thousand to four thousand, primarily in Canada and the United States. The Mammoth Jack is used primarily to breed huge draft mules for farming.

Sheep often come to see their guard donkey as their protector, running to her if they are frightened.

Spanish Jack

Spanish Jack is the popular term for a large donkey with refined, more elegant Spanish characteristics that is not large enough to be classified as a Mammoth Jack. Spanish Jacks are often used for breeding saddle mules.

Miniature Mediterranean

Miniature Mediterranean donkeys are the descendants of imports from Sicily and Sardinia, where they carried travelers and heavy loads, pulled small carts, powered grinding mills, and even produced milk for infants and the sick. Imports of these little donkeys began in 1929. The first registry was formed in 1958, although the true Mediterranean stock was bred to many other small donkeys already present in North America. Miniature donkeys are not rare, but breeders of quality minis can demand high prices.

Minis weigh 200 to 350 pounds and stand less than 36 inches at the withers. They are compact and well rounded but can vary in appearance from slender to draftlike in type. The legs should be straight. The tendency toward dwarfish characteristics should be avoided, especially in the size of the head and the shortness of the neck. Buyers need to be careful to look for physical defects when purchasing a mini, since there are some unscrupulous breeders. Miniature Mediterraneans are available in many colors and patterns, from light tan to dark black. White roan and spotted donkeys are also seen. Because of their size, miniature donkeys are generally a poor choice as livestock guardians. They can live twenty-five to thirty-five years.

Poitou

The extremely rare Poitou breed numbers only a few hundred worldwide. The Poitou is a horse-size donkey whose outstanding features include a huge head and ears and a long, thick, wavy coat that can form long mats or cords. Unlike other donkeys, the Poitou also grows a forelock and a long mane. Her hairy ears are so long they often flop to one side.

The Poitou, named for its native region in France, was once used to breed large work mules in France. Several Poitou donkeys have been exported to the United States, where they probably contributed to the Mammoth Jack, but mainly they were lost in the general donkey pool. Occasionally Mammoth Jack or American Standard donkeys will grow a longer, wavy hair coat as possible evidence of a Poitou ancestry.

Adopting a Wild Burro

The U.S. Bureau of Land Management's National Wild Horse and Burro Program makes wild horses and burros available for adoption in the United States. Approximately five hundred to a thousand burros are available for adoption each year. In order to adopt a burro you must be at least 18 years of age, have no convictions for inhumane treatment of animals, and demonstrate that you have appropriate facilities to care for a burro. There is a nominal fee. Adoption events are held in various locations around the country.

After one year of care and the submission of a signed statement from a veterinarian or another qualified person, you will receive a certificate of title for your burro. There are restrictions that prevent you from immediately disposing of a burro if it turns out to be unacceptable as a livestock guardian.

The wild burros that are offered for adoption are the descendants of escaped animals in the western states. The typical wild burro is 44 inches tall and weighs about 500 pounds. Jennies may be preferable, since a male will need to be gelded and kept from stock for 90 days. Take the time to observe the available burros. Choose a burro with a friendlier personality over one with good looks. Remember that a very young burro will not be a reliable guardian until she matures. A halter and a short lead line will be placed on your burro before you take possession of her, but don't expect much in the way of training. The burro will have received basic medical care, though, and you will receive a medical history.

Although your formerly wild burro will require taming, she can make an excellent livestock guardian. Do not immediately turn her loose in a large pasture, since you will have great difficulty catching her! A sturdy corral of 400 square feet with a 4½-foot fence is recommended. An acceptable shelter can be a three-sided lean-to.

Do not initially house your burro with any companions; isolation will speed up her interaction with you. Visit your burro several times a day, bringing her food and other treats and talking to her constantly. Slowly condition her to your presence by coming increasingly closer to her food each day. Eventually you will be able to stroke her neck, shoulders, and back gently, mimicking the mutual grooming of equines. You will need to train her to lead, to stand tied, to be groomed, and to have her feet trimmed.

Once she comes to you for treats, she can be turned loose with stock, following the same guidelines as with a thoroughly tamed donkey. Experienced burro adopters estimate that it will take a month of steady work to gain a burro's basic trust. Be cautious, and don't place yourself in a dangerous situation.

Introducing Your Guard Donkey to Stock

There are two schools of thought about training a donkey to become a livestock guardian. Some breeders advocate raising a young donkey with the stock she will eventually guard. Obviously, a donkey born and raised with stock will be thoroughly accustomed to them. Others believe that an adult donkey can be bonded to a flock over a period of weeks or months. In either case, it appears to be very important that the donkey not be socialized to dogs.

Agricultural agents and experienced users recommend that a donkey be used in small, open pastures with no more than two hundred to three hundred sheep. Dense vegetation and rough terrain will lessen the donkey's effectiveness. Some breeds of sheep or goats tend to scatter rather than flock, and this will also decrease the donkey's success. Use only one jack or jenny, or one jenny with a foal, per pasture. Pairs of donkeys generally spend their time together and do not stay with the stock. If you do have a jenny with a foal, remember that foals can be sexually mature as early as five to six months of age. At that time you can separate the mother and foal, leaving the foal with stock or selling it as a livestock guardian prospect.

There will be a period of adjustment for the donkey and stock as they become familiar with each other. Plan to obtain your donkey before your period of greatest predator pressure. Many users report a four- to six-week adjustment period for their donkey, but others report that it may take as long as six months or more for the donkey to be comfortably integrated with the stock.

At first the new donkey should be placed in a fenced area next to the stock she will eventually guard. Do not place her next to horses or other donkeys at this time or later when she is pastured with her stock. After a week or so, try tying the donkey in the pasture with the stock for short periods of time. If she is restless, you can groom her or feed her during this time.

As an alternative method, some experienced breeders advocate isolating the new donkey completely from other animals for a period of one to two weeks. She will be very lonely and probably very loud at times. This method may sound a bit cruel, but it might speed up her acceptance of her new herd mates, because after this isolation period she will be anxious for companions. You can visit her during this time, and it is a good opportunity to practice your donkey-handling skills. After the

Allow the donkey and stock to become acquainted with each other over a fence for at least a week or so before turning them out together.

period of isolation, place her in a paddock or pen next to her new stock for a few days.

After a week to ten days, lead the donkey into the stock pasture. Let the animals smell each other. Spend time grooming the donkey or letting her eat. If the donkey can be safely tied, you can leave her for short periods in the pasture with the stock. If possible, turn the donkey loose with the stock in a small area where you can observe her behavior. You don't want to overwhelm and stress her with too many animals or too small an area. Some donkeys are excellent guardians with stock in the field but find it difficult to be confined with them in a small paddock or barn. Be vigilant the first few times you do this, just to be safe.

If all is going well after a few days, you can leave the donkey loose in the pasture. Do not leave her loose at night or when you will be unable to monitor her and the stock for several hours. Be observant for several days, taking care that the donkey and the stock are accepting each other and that there are no problems at feeding times. Some donkeys behave too roughly with sheep or goats, treating them like other donkeys. Donkeys may also chase the sheep, bite at the ears or wool, and protect food or water sources. If this occurs, try penning the donkey near the stock again for a few more days to allow her to become more accustomed to her new herd mates. You can also try the isolation method again, but repeated failures probably indicate that this donkey will not make a good guardian. You really can't train a donkey like a dog.

CHANGES IN PASTURE OR STOCK

Donkeys don't generally have any trouble changing pastures or fields with their stock. Your biggest concern in rotational grazing systems will probably be that the new pasture may be too rich for your donkey. Donkeys are definitely at risk for both obesity and founder from lush pastures. You also need to introduce donkeys slowly to pasture when they are coming off dry feeds such as hay. You may need to use an anti-grazing muzzle to limit her access to grazing.

A problematic time might be when you introduce a ram or buck into your flock for breeding. The donkey's unfamiliarity with the new animal, combined with the ram or buck's behavior to the females, may cause the donkey to see him as an intruder.

Fitting a Donkey Halter

Horse halters do not fit donkeys very well. Some manufacturers do make halters specifically created for donkeys, but it can take some effort to find them. The Internet may be the best source.

It is possible to adapt a horse halter for a donkey. Generally the problem is that if the noseband fits, the halter is too long, or if the halter is the right length, the noseband is too tight. To remedy this situation, buy a halter with a noseband that fits correctly and then punch additional holes in the crown piece to pull the halter up on the donkey's face.

The noseband of the halter should be loose enough to allow you to place three fingers between your donkey's chin and the halter. The noseband should rest halfway between her eyes and her nostrils. Donkeys can be a bit sensitive about their ears, so it is best to buckle the crown piece behind the ears rather than dragging the halter up and over the ears.

For safety's sake, do not leave a halter on an unattended donkey. The donkey could easily catch the halter on a fence or other obstacle or become entangled when she scratches her head with her hind hoof. Check the halter for fit during the period of heavy coat growth in the winter; the halter can cause sores if it becomes too tight.

10
TAKING CARE
OF
YOUR DONKEY

Though generally hardy and healthy, donkeys need the same routine care as horses: annual immunizations, regular deworming and parasite control, hoof trimming, and dental care. While your donkey can live quite happily outdoors and graze on the same pasture as your livestock, she does require shelter from harsh weather, especially heavy rain or snow.

WHAT TO FEED YOUR DONKEY

It is very easy to overfeed a donkey, which can lead to serious consequences. A fat donkey will develop a thick crest or roll on her neck, as well as pads of fat on her barrel and on her hips. Donkeys evolved to graze arid or coarse pastures. They require only a half to one acre of growing pasture or rotating pasture per month. During the winter grass, timothy, or mixed hay is most suitable and can be given at the rate of two to four flakes per day for a Standard donkey. High-quality legume or alfalfa hays may be too rich for donkeys, and silage may have a too high a protein level for them. When not on pasture a donkey will need about a third of a bale or 12 pounds of hay every day, depending on her size. In the spring, avoid a rapid change from hay to rich, fresh spring grass to avoid colic. Mildew or mold on hay is poisonous to all equines.

Grain or other concentrates are seldom needed. In fact, adult donkeys do not need any grain at all if they are eating good pasture or hay. Some high-energy feeds for lambs and calves are definitely too rich for donkeys and may cause founder, potentially crippling your donkey or even causing death. If you wish to feed some grain, about one cup of plain rolled oats is sufficient.

Dried sugar beet pulp is safe for your donkey in small amounts, although it should never replace hay

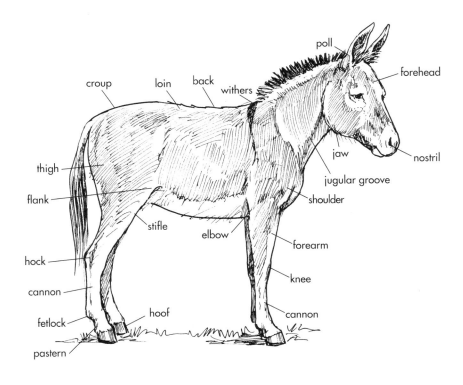

in her diet. Avoid the sugared or molasses formulas, as they are too rich for most donkeys. Beet pulp should always be soaked before being fed, as dried beet pulp can expand in the stomach and cause serious digestive problems.

Donkeys need access to trace-mineral salt formulated for equines (as opposed to the pure white salt eaten by sheep) and larger quantities of fresh water than small ruminants.

Perhaps the most important consideration is to remember that donkeys are not ruminants, and therefore their digestive tract cannot always deal with the diets of sheep, goats, or cattle. Counter to folklore, donkeys are just as susceptible to overeating, colic, and founder as horses.

PROVIDING SHELTER

Donkeys need shelter from inclement weather, which may be a problem if your shelters are only sheep or goat sized. Donkeys are natives of the desert and generally dislike rain, wind, and wet ground, although they will adapt to cold or humid conditions.

Donkeys have longer coats than horses but lack a protective undercoat. While rain runs off a horse, it soaks the coat of a donkey. This makes them more susceptible to susceptible to skin problems.

Additive Caution

Feed additives containing anabolic agents or urea such as monensin (trade name Rumensin), lasalocid (trade name Bovatec), or decoquinate (trade name Deccox) are extremely poisonous to horses and donkeys. These medications are often used to prevent coccidiosis in sheep and goats and are fed to the pregnant females the month before birth. If you use these preventives, be aware that even small amounts are toxic to equines.

In a warm or temperate climate your donkey will be comfortable with a three-sided shelter to provide shade and a place to get out of the wind or occasional rain. She will also seek the darkness to avoid flies. Orient the shelter away from the prevailing winds.

In a cold climate your donkey will need to be kept in a barn or snug shelter during times of subfreezing temperatures, strong winter winds, or freezing rain and snow. A hard wind can blow away the insulating air in her coat, putting her at risk of wind chill. Straw or shavings both make good bedding for donkeys, but note that some donkeys will eat straw. Also, beware of black walnut shavings, since they are toxic to donkeys and will cause founder if ingested.

Some donkeys enjoy snow during calm weather, but melting snow will soak her coat, so any snow that may accumulate on your donkey's back should be brushed off before she is returned to the barn. Donkeys may also have trouble walking on icy ground during severe weather.

Fencing is not usually a problem for donkeys. Good fencing and an outside scare wire will help them do their job as guardians. Woven wire, electric, or high-tensile wire fencing all work well with donkeys. They learn quickly to respect electric fencing.

Routine Health Care

Donkeys need yearly vaccinations, which may include those for Eastern and Western equine encephalitis, tetanus, influenza, West Nile fever, rhinopneumonia, strangles, and rabies, depending on the area where you live. They also need to be routinely treated for worms and other parasites. Most of the dewormers, vaccines, and topical insecticides used on donkeys have been tested only on horses, though it is assumed that they will be safe for their cousins. Occasionally, however, there is news

Since donkeys lack a protective undercoat, your guard donkey will need shelter from rain, snow, and wind.

of adverse or allergic reactions in donkeys to horse medications. Consult your veterinarian for specific information on new medications and a schedule of treatment for your donkey.

Most states now require equines to have an annual Coggins test, which checks for equine infectious anemia (EIA). In some states it is required only when you are moving equines off your property, while other states that have serious problems with the disease mandate that all equines be tested yearly. There is no vaccine and no cure for EIA. Infected equines usually succumb rapidly; those rare animals that recover become dangerous carriers. The symptoms can be very general, which can make the disease difficult to diagnose without testing. Equines with the acute form die within a month, although chronic forms may allow the horse to linger for a year or so. The disease is spread through biting horseflies.

Testing is the only prevention tool we have for EIA, and it has dramatically reduced the number of fatalities from this disease. Without this inexpensive testing, you risk bringing home an infected animal. Although EIA is found in most states, in recent years the largest numbers of cases have been found in Oklahoma, Louisiana, Missouri, Texas, Mississippi, Alabama, Florida, Iowa, and Nebraska.

HOOF CARE AND TRIMMING

The donkey's hoof is actually more durable than the average horse's hoof and in most cases donkeys do not require shoes. However, they can have foot problems in wet environments or when they are fed overly rich diets. You must pay attention to several potential problems associated with or exacerbated by wet conditions or overfeeding, including thrush, abscesses, white line disease (also known as seedy toe), and laminitis (also known as founder).

The hoof should be trimmed to match the pastern angle found on the individual animal. In comparison to horses' feet, donkey feet are generally more upright and "boxier," or smaller and rounder, and the sole is more U-shaped. Donkeys often grow more hoof from the heel area, which can add to the "up on the toe" appearance. The American Donkey and Mule Society offers a wonderful informational packet on how to trim donkey hooves, available for a small cost with an e-mail request. (See Resources.) This packet can be very helpful to your farrier if he is unfamiliar with donkeys.

It may be difficult to find a farrier, especially one who is familiar with donkeys and understands their unique behavior as well as their feet. A good farrier is worth his weight in gold, so treat him with respect and appreciation. Besides being present at your appointment to assist him (and paying him promptly!), the most valuable thing you can do is to practice picking up your donkey's feet regularly between your farrier's visits. Clean them out with your hoof pick, tap them with the pick a few times, and even ask your farrier to show you how to rasp them a little.

All of this will make your donkey more comfortable with the farrier's work. Praise good behavior. Donkeys may need their hooves trimmed as often as every two months, depending on their hoof growth, the season, and your ground. Regular foot trimming will be a reality of your donkey's long life, and there is no reason why it cannot be a routine chore rather than an unpleasant battle.

Regular cleaning of your donkey's hooves will not only make the farrier's job easier, but will alert you to potential problems.

Since most donkeys are shorter than horses, be careful not to lift the donkey's foot as high as you would with a horse. To pick up a foot, stand next to the donkey's shoulder and face the rear. Run your hand down her leg and lift the hoof from the fetlock. When the donkey lifts her foot, place your hand under it for support. Put it down gently when you have finished.

Regular attention to your donkey's feet will go far in preventing problems. Picking out the accumulated manure, dirt, and rocks regularly will also allow you to learn what a healthy hoof looks like and detect potential problems early. Whenever you see any signs of lameness or tenderness in your donkey, pick up her feet and clean them out. The problem may be as simple as a small rock lodged in the crevice of her frog.

CARE OF TEETH

Equine teeth grow continuously and are ground down by grazing and chewing. Equines have two types of teeth: the incisors in front that bite off grass and the molars in back that grind and chew food. Many horses and most donkeys also have small, pointed canine teeth that appear in the space between the front and back teeth, as well as up to four wolf teeth that grow just in front of the premolars. Wolf teeth are usually small and shallow-rooted, but they can cause trouble for animals that are driven or ridden with a bit. A veterinarian or equine dentist can remove these teeth for you if they are a problem.

A veterinarian should check your donkey's teeth every year; this is especially important for older donkeys. If sharp edges, called points, develop, they should be smoothed down with a dental rasp, a process called floating. Older donkeys can also lose teeth. Signs of mouth problems include drooling, tilting or shaking the head while chewing, and dropping quids, or wads of partially chewed food, out of the mouth. Quids can sometimes be found on the ground.

EXTERNAL PARASITES

A donkey's hair coat sheds much later in the summer than does that of the horse, which serves to protect the donkey from some parasites. Nevertheless, donkeys can be bothered by skin parasites such as fleas, flies, gnats, mites, mosquitoes, and ticks.

Besides being annoying, biting flies and mosquitoes can transmit disease. Sprays and wipe-on products for horses are generally safe for donkeys. Do not use other livestock products on donkeys. Since flies develop some resistance to the common products, controlling these pests means eliminating their breeding sites. Mesh masks made for horses can offer some respite from flies and gnats that bother the eyes. If you use a mask, make sure it fits properly and check it frequently to ensure that it is not rubbing off hair or causing sores.

Lice are most common in the winter, when animals have longer hair and are often housed more closely with other livestock or poultry. Look for the donkey rubbing her coat, especially at the tail head. Pediculicide powders can be difficult to use on the long coat of a donkey. A 4 percent solution of permethrin can be used as a pediculicide, but other pour-on and injectable products are not approved for treating equines.

Mange mites can cause dermatitis and infections. If you suspect your donkey has mange, consult your veterinarian for a diagnosis via skin scrapings, and treatment options.

INTERNAL PARASITES

Donkeys are affected by the worm species typical for all equines. The most common ones include large strongyles, small strongyles, ascarids, tapeworms, bots, threadworms, pinworms, hairworms, stomach worms, and lungworms. In most cases, although they are often not tested on donkeys, the same anthelmintic dewormers that are appropriate for horses will work for donkeys. Good sanitation and pasture management are also essential in controlling internal parasites. Failure to control worms will lead to damage to the donkey's digestive tract and eventually death.

Your veterinarian can do a fecal egg count from a manure sample to help you determine the correct parasite control program for your donkey. Deworming medications are usually given every eight weeks,

Deworming Your Donkey

Different medications kill different worms. Ivermectin is used for ascarids, large and small strongyles, pinworms, bots, lungworms, stomach worms, threadworms, and hairworms. Fenbendazole, oxibendazole, and pyrantel are effective against ascarids, large and small strongyles, and pinworms. Problems have been reported when moxidectin (trade name Quest) has been used for donkeys.

You will need to know your donkey's weight to administer deworming medications. Horse weight tapes (available at feed or horse supply stores) should give you a good estimate, unless your donkey is especially stocky or short-legged. You can also ask your donkey's breeder, your veterinarian, or a person familiar with donkeys to give you a rough approximation.

To administer paste dewormers to a donkey, first preset the dosage amount on the tube. Tie your donkey or have another person hold her. Make sure her mouth is empty, and then insert the syringe into the corner of her mouth at a 45-degree angle. Push the syringe into her mouth and depress the plunger. Lift the donkey's head to encourage her to swallow the paste. If she resists swallowing it, try offering her a small amount of loose grain in a bucket as soon as you administer the dose.

and they should be rotated to decrease the occurrence of resistance in parasites. There are also daily dewormers for use where donkeys are continuously exposed to contamination. Do not use injectable or pour-on dewormers without professional advice. Consult your veterinarian for tapeworm treatment.

Lungworms *(Dictyocaulus arnfieldi)* are more common in donkeys than in horses, and donkeys are the prime source of lungworm pasture contamination for horses and other animals. Lungworm is associated with severe coughing in horses, but donkeys may show very few symptoms. If you own horses, be sure to prevent lungworm infestation in your donkey.

Bot flies are active in the late summer and early fall, gluing their tiny eggs to body hairs of equines, most commonly on the forelegs, but also on the outside of the legs, the mane, the flanks, the long hairs beneath the jaw, and the hairs on the upper and lower lips. Newly hatched bot larvae fall to the ground and enter the host animal's mouth as she grazes. The larvae spend about three weeks in the soft tissue of the lips, gums, and tongue. The bots then migrate to the stomach or small intestine, where they can damage the organ's lining, interfere with the passage of food, or cause other gastrointestinal problems. They must be treated internally. In temperate areas one dose of ivermectin, two weeks after the first frost, will treat that season's bots. You must use ivermectin for this worm even if you use different or daily dewormers for other parasites.

GROOMING YOUR DONKEY

Donkeys enjoy being groomed, and grooming helps build the bond between you and your donkey. Always approach a donkey in a calm and relaxed way. Talk quietly to her and avoid sudden movements. Brush the hair in the direction it grows with a stiff bristle brush.

Donkeys grow a heavy winter coat, which they typically shed about two months later than horses in the same climate. In the spring especially they need some assistance with grooming to shed that heavy coat and feel more comfortable. Some owners clip their donkeys to speed up this process and make them more comfortable in hot weather. A shedding blade can help remove the winter coat. Avoid clipping unless it is necessary for especially humid climates, since it removes the donkey's protection from harsh or wet weather.

It is critical to keep your donkey's coat free of mats, debris, and dried manure or mud in the winter. Since donkeys lack an undercoat, the air pockets between their hairs are their only insulation, and they need to be able to fluff up their coats. For this same reason, avoid grooming in winter on very cold or cloudy days so that you don't slick down the coat.

Specific Health Concerns

Donkeys are noted for their hardiness and long life span, but they are susceptible to many of the same ailments as horses, and precautions should be taken to prevent the following problems.

COLIC

Colic is a painful symptom of trouble somewhere in the digestive tract. Although some donkeys will show the same signs of colic as horses do, many will not. Donkeys are very stoic, and often the only sign of colic in a donkey is described as overall "dullness" and an unwillingness to eat. Other signs of colic are rolling, pawing the ground, pawing at the belly, increased respiration, sweating, a dark red color in the gums or inside the eyelids, and a lack of fresh manure.

If you see these signs or sense that something is not right with your donkey, you must call the veterinarian. Depending on the type of colic, the veterinarian may pass a stomach tube, give mineral oil by tube or enema, or inject painkillers or antibiotics. Surgery may be required to save an animal that has developed a twisted gut.

Colic can be caused by several things, most of them to do with feeding. Sudden changes in feed such as going from grass to hay or from hay to rich grass, eating too much grain, ingesting bedding materials, having dry sugar beet pulp, or even eating bizarre things like plastic bags can all cause colic or intestinal difficulties. Just like horses, donkeys can develop sand colic if they ingest too much sand while eating hay or grass; there are sand colic supplements that can be added to the donkey's food if this is a problem.

Lack of fresh drinking water during cold weather can also cause colic. Large numbers of internal parasites can cause colic, as can poisonous plants or even stress. Donkeys like routine, so any sudden changes can cause colic. Closely observe your donkey for twenty-four hours after transporting her anywhere, as this can be a stressor.

Prevention is the best cure for colic, and early intervention increases the chances of a successful outcome. Try to observe your donkey enough to know her usual habits — how much she eats, how much manure she makes, and her usual behavior. Remember that donkeys evolved in arid areas with poor forage, so rich foods, such as spring grass, alfalfa hay, and grain, are especially dangerous.

Make sure that fresh, open water is available to your donkey at all times, and particularly in winter weather. Follow a regular deworming program to prevent a dangerous buildup of nasty parasites. Finally, have your veterinarian check your donkey's teeth once a year or whenever you observe difficulties. Problems in the mouth or with the teeth can prevent the donkey from chewing her food properly, which can lead to colic.

FOOT PROBLEMS

While donkeys typically have strong, healthy hooves, you must routinely check to make sure that your donkey's feet are sound. Injuries, improper diet, and overly wet conditions can bring on serious problems that require medical attention.

Abscesses

A foot abscess is the most common cause of lameness in a donkey. Abscesses form when a nail, a piece of gravel, or any sharp object penetrates the most vulnerable area of the hoof: the "white line" between the sole and the hoof wall. The puncture introduces bacteria, and as the infection grows, pus builds up and causes serious pain. Long periods of standing on wet ground open up the hoof to infection, as do alternating periods of very dry and very wet conditions.

A farrier or veterinarian can often cut open small abscesses so that they will drain. You may need to soak the foot for a few days, apply antiseptics, provide antibiotics, and keep the donkey off wet or dirty ground. Duct tape is helpful for creating a waterproof bandage over absorbent packing, if this is necessary. Make sure your donkey's tetanus inoculation is up to date as well.

If gravel or other matter works its way up into the white line, the infection may not break open until it reaches the coronary band at the top of the hoof. In such a case you should consult a veterinarian to help you prevent more serious consequences.

Laminitis

Laminitis (sometimes called founder) is a serious hoof disease for horses and donkeys. The coffin or toe bone is attached to the hoof wall by the laminae, thin plates of soft vascular tissue. Laminitis is an inflammatory disease of the laminae. The front hooves are most commonly affected by laminitis, although the hind feet are sometimes affected as well. Not all horses that develop laminitis will founder, but all horses that founder first experience laminitis. When a horse founders, the coffin bone rotates and sinks down toward the sole of the foot. Severe founder has a very poor prognosis.

The symptoms of laminitis or founder can be seen in the donkey's stance. She may be reluctant to move, shifting weight from front foot to front foot or rocking back on her heels with her front feet extended. This is a veterinary emergency. While you wait for help, remove all food except grass hay and let your donkey rest on very soft footing. Your veterinarian may recommend that you give her anti-inflammatory, pain-relieving medications such as phenylbutazone (commonly called bute) or aspirin. The vet may recommend supportive shoes. Severe cases are crippling, and euthanasia may need to be considered.

There are several potential causes of laminitis. Overly rich food is the leading cause. Obesity seems to predispose a donkey to founder. Black walnut shavings, poor foot care, and illness or injury can also contribute to this problem.

Thrush

Thrush is a bacterial infection of the frog. Thrush is anaerobic, which means it can't survive in the presence of oxygen. It is easy to diagnose thrush because it produces a distinctive and foul odor that you will notice when you pick out the hooves. The frog itself may appear soft and ragged and have a black discharge. After the affected area is trimmed away, you can easily treat the infection with a thrush remedy available at feed or farm supply stores. Wet, muddy conditions and hooves that are not cleaned out regularly allow thrush to thrive. If thrush is ignored, the frog may ulcerate and the infection may become difficult to treat successfully.

White Line Disease

White line disease affects the area between the hoof wall and the sole. The white line is the nonpigmented inner layer of the hoof wall. White line disease is more common in hot, humid areas. Prevention includes regular hoof care and avoidance of muddy areas, concentrated manure, and dirty bedding. Chronic laminitis may contribute to this problem.

In its early stages the disease is known as seedy toe. At this stage a sharp-eyed farrier can often detect a small area of powdery, weak, or crumbly tissue at the white line in the center of the toe. The donkey will usually not show any signs of lameness at this point. Your farrier can test the area for soreness with hoof

A foundered donkey will be reluctant to move and she may rock back to carry more of her weight on her hind feet.

testers or tap on the hoof with a hammer to detect a hollow sound. The affected area can be removed and the foot treated with Merthiolate, Betadine ointment, or 2 percent iodine to help dry the area. The foot can then be packed with gauze and covered with duct tape to hold the gauze in place. The damaged tissue should be cleaned out every two weeks or so.

More serious white line disease may require removal of the outer hoof wall to expose the diseased tissue. The hoof must also be monitored for abscesses.

RAIN ROT

Rain rot or rain scald is a fungal infection that is exacerbated by wet weather. The symptoms are crusty, scabby patches and hair loss. It takes persistent effort to rid your donkey of rain rot. She will need regular shampoos with Betadine or iodine shampoo. The hair may fall out in large patches, but it will grow back. If the infection becomes severe, your donkey may need antibiotics.

RESPIRATORY DISEASE

Respiratory disease affects donkeys as it does horses, although the symptoms and duration may differ. It may be either viral or bacterial in nature. In addition to looking "dull" or showing a lack of interest in eating, the donkey may exhibit symptoms such as a high temperature, discharge from the nose, swelling under the jaw or in the throat, difficulty breathing, increased respiration, and coughing. Mild symptoms, such as a bit of a runny nose, may go away in a few days, but if they don't or if you see more serious symptoms, you should call a veterinarian. Your donkey may need antibiotics. A good immunization program will prevent most serious respiratory diseases.

Coughing can also be caused by dusty feed and bedding or allergies. Donkeys and other equines need good ventilation. Even in winter, stables or barns should not be shut up tight.

Solving Problem Situations

Most people who decide to employ a livestock guard donkey against their predator problems will encounter one or more of these situations or problems. Fortunately, experienced users have developed several techniques and suggestions for dealing with these common dilemmas.

PROBLEM: Aggression toward Humans

SOLUTION: Donkeys do not generally show aggression toward humans. They are usually friendly and docile unless badly frightened or mistreated. Sudden changes in temperament often indicate that a donkey is in pain or ill. The greatest difficulty in handling donkeys is with the intact jack, which should never be used as a livestock guardian under any condition.

Feeding treats can cause a donkey to become pushy, so to avoid conditioning your donkey to look for treats in your hands, offer her treats from her feed tub while you pet or groom her. If your donkey does attempt to bite, never hand-feed her again. Yell a loud *no* immediately, then walk away and ignore her. Be consistent. Most donkeys want your friendly attention.

Donkeys that attempt to run away or kick you are exhibiting avoidance of frightening events or have learned that you don't always insist on good behavior. Calm insistence is the best method with a donkey. You can correct bad behavior, but only if you stop the donkey in the act, not later. Ignore the bad behavior and reward good behavior. Forceful punishment will not help. In fact, violence or anger gets you absolutely nowhere with a donkey. If you have frightened your donkey in the past, it will take time to regain her trust and reestablish your leadership.

More serious problems require an experienced handler. The techniques of natural horsemanship and round pen work, in particular, are very successful with donkeys.

PROBLEM: Aggression toward Stock

SOLUTION: This is a major problem area for guard donkeys. Try to avoid situations in which your donkey becomes aggressive in defense of her food. Feed her every time you give grain or hay to the stock. Some owners suggest that you feed the donkey first, from a separate pan or trough, to minimize potential conflicts over food. This is more of a problem with sheep or goats than with cattle.

Some jennies can become aggressive during estrus and may injure young or small stock. You can remove her temporarily or even consider spaying, if she is a good guardian at other times. The good news is that jennies do not generally show estrus throughout the winter.

PROBLEM: Aggression toward Family Dogs

SOLUTION: Donkeys are not usually placed together with livestock guard dogs to protect stock. Some owners who started with a donkey as a guardian have found it necessary to include dogs in their protection program. If the donkey accepts the dog or dogs, they can all be left together. In this case, however, the donkey may become less aggressive toward predators.

If your donkey is aggressive toward canines, you may have trouble using herding dogs or farm dogs around him. A donkey can easily kill a dog, so please exercise caution. You may need to remove the donkey when you use herding dogs. Some donkeys will learn to accept the farm's dogs over time and remain alert to strange canines; however, it is not advisable to allow your donkey and dogs to become overly friendly toward each other. Do not allow dogs to tease the donkey.

PROBLEM: Birthing Season Issues

SOLUTION: You may need to remove the donkey from the stock during birthing times to prevent accidents, especially in confinement situations. The birthing process seems to unsettle some donkeys. The newborn is unfamiliar to them and may be seen as an intruder. Donkeys have been known to pick up and shake lambs. If they are too possessive or interested in the newborn, they may also interfere with the mother and her offspring by separating them.

If it is your practice to remove your guard animal during birthing season, try to keep your donkey very near her flock during this time, and carefully reintroduce them when the animals return to pasture, observing her reaction to the babies. This method works well in situations where you lamb or calve in close confinement. Be very cautious about confining your donkey in a very small space with her stock.

If you lamb or calve out on pasture or range, you should still exercise caution during the donkey's first birthing season. If you find that after a year's experience you must still separate the donkey and her flock, then it will be preferable to find a calmer guardian donkey that does not need to be removed from the flock, especially if this is a time of high predator threat to your animals.

PROBLEM: Excessive Braying

SOLUTION: The donkey's raspy bray is famously loud and may irritate your neighbors. Donkeys bray when they are hungry and are expecting a meal. They bray if they are lonely and lack companions, so you can expect more braying in the beginning after bringing your donkey home. Jacks bray a great deal, and this is another good reason for choosing a gelding or jenny. Geldings may still bray more than jennies. Young donkeys bray more than older ones.

Be careful not to condition your donkey to bray every time she sees you in the hope of a treat or a friendly visit. The best prevention of this type of braying is to feed the donkey at the same time every day, instead of at irregular times, and not to pass out treats every time you visit the pasture or paddock. Feeding the larger portion of your donkey's ration at night will keep her content until your morning chores and lessen the early morning braying that might really annoy your neighbors. Do not reward braying by visits or food. Donkeys learn very quickly, and this pattern can be established almost overnight.

PROBLEM: **Hard to Catch**

SOLUTION: Building a catch pen or small corral adjacent to or around your donkey's shelter in her pasture will make it easier to catch her if she won't come to you willingly. In addition, it will serve as a good place to secure her when you need to work with your stock or need to isolate her for some other reason. This catch pen should be made of substantial materials, like wood or pipe, rather than something she could run through and potentially injure herself on as she tries to return to her herd mates.

If your donkey is hard to catch in a large field, entice her into the catch pen with something good to eat like carrot or apple pieces. Build positive experiences by walking up to her in closer quarters and scratching her withers. Do not pat her, since she may interpret it as swatting or hitting. It may take several days for her to learn to stand while you do this. Practice several times without attempting to halter her. Positive behaviors should be praised immediately. Stubborn or scared donkeys can be won over with this approach. Truly aggressive or wild donkeys should be handled with great care.

PROBLEM: **Chewing on Wood**

SOLUTION: Donkeys like to chew wood, and they can quickly do a great deal of damage. They will even chew pressure-treated lumber. If your donkey is whittling away at an important post, try anti-chew paint or metal flashing.

PROBLEM: **Failure to Guard**

SOLUTION: If your donkey fails to bond to the stock in a few weeks, you may need to give her a few months more to adapt to her new situation and herd mates. Experienced equine owners believe that any equine needs to spend a year at a new place before she is fully comfortable with the changes in seasonal routines. However, if your donkey still does not guard well or totally ignores predators, she may need to be sold. You should be prepared to try another donkey until you find one that works well. This is not failure on your part, but rather a lack of aggressive instinct in your donkey.

Riding or Driving Your Donkey

If you purchased your donkey primarily as a guardian, you may not be interested in doing other things with her. However, using your donkey for other activities will not reduce her effectiveness against predators as long as she spends most of her time with her stock. Standard donkeys are very strong for their size, and once they are mature they are quite capable of being trained to pull a cart or even for light riding, though donkeys are not as comfortable to ride as horses due to their short strides.

Since donkeys lack true withers, you can prevent the saddle from slipping forward by using a crupper under the tail. Most horse saddles will not fit donkeys. Look for an Australian or McClellan cavalry-style saddle, as these two models seem to be more adaptable to fitting a donkey comfortably. The girth may need to be relocated so that it does not irritate the donkey.

APPENDIX AND RESOURCES

APPENDIX A:
Predator Information

These groups conduct research and disseminate information concerning predator control and livestock management. There is a great deal of information available on their Web sites and through their published research.

Defenders of Wildlife
800-385-9712
www.defenders.org
"Defenders of Wildlife is dedicated to the protection of all native wild animals and plants in their natural communities. . . . Defenders of Wildlife also advocates new approaches to wildlife conservation that will help keep species from becoming endangered."

The Bailey Wildlife Foundation Wolf Conservation Trust pays livestock owners for losses to wolf predation. Proactive Carnivore Conservation Fund promotes innovative approaches to prevent conflict between endangered predators and humans. (Click on "Wildlife" and then "Wolves" to find information on these programs.)

Grupo Lobo
www.timberwolfinformation.org/info/world/grupolobo.htm
Grupo Lobo is an independent, non-profit organization dedicated to the conservation of the Iberian wolf and its habitat.

International Wolf Center, Depredation Information
218-365-4695
www.wolf.org/wolves/learn/intermed/inter_mgmt/depredation.asp
"The International Wolf Center advances the survival of wolf populations by teaching about wolves, their relationship to wild lands and the human role in their future."

Internet Center for Wildlife Damage Management
www.icwdm.org
"The Internet Center for Wildlife Damage Management (ICWDM) attempts to consolidate existing and future information on integrated pest management (IPM) in wildlife damage management. Its goal is to increase adoption of IPM practices in wildlife damage management by centralizing resources." Provides many links and publications.

KORA
www.kora.ch
KORA coordinates research for carnivore conservation in Switzerland.

Large Carnivore Initiative for Europe
www.lcie.org
LCIE "seeks to maintain and restore, in coexistence with people, viable populations of large carnivores as an integral part of ecosystems and landscapes across Europe."

The Mountain Lion Foundation
800-319-7621
www.mountainlion.org
"The Mountain Lion Foundation is a nonprofit conservation and education organization dedicated to protecting the mountain lion and their wild habitat to ensure that our wildlife heritage endures for future generations."

National Agricultural Statistics Service
800-727-9540
www.nass.usda.gov
"The National Agricultural Statistics Service provides timely, accurate, and useful statistics in service to U.S. agriculture."

National Wildlife Research Center
970-266-6000
www.aphis.usda.gov/ws/nwrc
"The U.S. Department of Agriculture's National Wildlife Research Center (NWRC) is the federal institution devoted to resolving problems caused by the interaction of wild animals and society."

Predator Conservation Alliance
406-587-3389
www.predatorconservation.org
The Predator Conservation Alliance promotes Predator Friendly certification, which entitles you to use the official Predator Friendly logo to market your products to individuals and other consumer markets.

U.S. Animal and Plant Health Inspection Service (APHIS) Wildlife Services
www.aphis.usda.gov/wildlife_damage/
"Provides Federal leadership and expertise to resolve wildlife conflicts and create a balance that allows people and wildlife to coexist peacefully."

Wild Farm Alliance
831-761-8408
www.wildfarmalliance.org
"The Wild Farm Alliance was established by a national group of wildlands proponents and ecological farming advocates who share a concern for the land and its wild and human inhabitants."

APPENDIX B:
Predator Friendly Logo Program

Many consumers are supportive of positive environmental efforts. Awareness of and demand for the Predator Friendly logo is growing, and its use often allows farmers and livestock producers to charge premium prices for their products. The following products can be certified Predator Friendly:

- Raw wool, felted wool, wool yarn, wool thread, woolen clothing and other goods
- Animal fibers, yarn, and goods made from Angora rabbits, Angora goats, bison, dogs, alpaca, guanaco, vicuna, donkeys, llamas, camels, and yaks
- Breeding stock
- Meats: beef, pork, bison, lamb, chicken, goat, turkey, duck, goose, emu, ostrich, and game birds
- Leather goods: sheepskin, cowhide, emu, and ostrich
- Dairy products: milk, cheese, yoghurt, sour cream, butter, and eggs
- Honey
- Soaps and lotions from animal by-products
- Horses, mules, and donkeys

Predator Friendly Practices

The Predator Friendly logo program recommends:

- Use of guard animals
- Fencing out predators while leaving corridors for wildlife to pass through
- Frequent appearance of humans in pastures
- Use of fladry and other devices intended to frighten predators off
- Grazing cattle with sheep, goats, and calves
- Use of birthing sheds and small secure fenced lots for lambing and calving
- Timing pasture use with periods of low predator pressure
- Timing calving and lambing to avoid periods of high predator pressure
- Prompt removal of carcasses
- Investigation of livestock deaths to determine if a predator was responsible
- Recording all predator death (time and location) to help plan future strategies

Contact the Predator Conservation Alliance (406-587-3389 or *www.predatorconservation.org*) for a copy of the certification criteria and affidavit. By signing the affidavit, you agree not to kill or ask others to kill native predators on your property for a period of one year. Certification, which costs $45, entitles you to use the official Predator Friendly logo to market your products to individuals and commercial markets such as restaurants, grocery stories, and Web-based retail sites. You can sell directly to consumers through the Web site *www.predatorfriendly.com* and you will receive access to marketing tools and identified retail markets in your region. You also will receive the biannual newsletter *Wild Guardian: A Field Journal for Coexisting with Predators*.

APPENDIX C:
Canine Health Registries

Canine Eye Registration Foundation (CERF)
217-693-4800
www.vmdb.org/cerf.html
"Dedicated to the elimination of heritable eye disease in purebred dogs through registration and research."

Orthopedic Foundation for Animals (OFA)
573-442-0418
www.offa.org
Mission: "To collate and disseminate information concerning orthopedic and genetic diseases of animals. To advise, encourage and establish control programs to lower the incidence of orthopedic and genetic diseases. To encourage and finance research in orthopedic and genetic disease in animals."

University of Pennsylvania Hip Improvement Program (Penn-HIP)
www.pennhip.org
"PennHIP's mission is to develop and apply evidence based technology to direct appropriate breeding strategies aimed at reducing in frequency and severity the osteoarthritis of canine hip dysplasia. The beneficiaries of this effort will be the many dogs who suffer with this controllable genetic disease and, of course, the dogs' owners."

APPENDIX D: Transporting Animals across State Lines

Center for Epidemiology and Animal Health
www.aphis.usda.gov/vs/ceah/
800-545-8732
Each state has separate requirements for animals entering or passing through their state. The Voice Response Service (VRS) of the Center for Epidemiology and Animal Health is available to individuals wanting to know which health documents and tests are required to transport livestock or pets to another state. The VRS is a 24-hour, 365-day-a-year access system that allows users to retrieve information about transporting animals. The retrieval system is accessed by simply dialing the toll-free number on a touch tone phone.

Resources

Suppliers

Anatolian Shepherd Dogs International
www.anatoliandog.org/anatolianlgd-sign.htm
Working livestock guardian signs

Carol's Custom Creations
Arco, Idaho
208-527-8538
fladry@ida.net
Fladry flags

Geotek Inc.
Stewartville, Montana
800-533-1680
www.geotekinc.com
Heavy-duty fiberglass fencing

The Mountain Lion Foundation
Sacramento, California
800-319-7621
www.mountainlion.org
Designs for mountain lion-proof housing for livestock and pets

Premier 1 Supplies
Washington, Iowa
800-282-6631
www.premier1supplies.com
Fencing, netting, sheep and goat supplies

Quality Llama Products & Alternative Livestock Supply
Lebanon, Oregon
800-638-4689
www.llamaproducts.com
Llama and donkey halters

General Reading: Books & Articles

Damerow, Gail. *Fences for Pasture and Garden*. (Storey, 1992)

Fytche, Eugene L. *Wild Predators? Not in My Backyard!* (Baird O'Keefe, 2003)

Guthrey, F. and S. L. Beasom. "Effects of predator control on Angora goat survival in South Texas" in *Journal of Range Management*. (31:168, 1978) Archives on the University of Arizona Library
www.library.arizona.edu

Kenkins, David. "Guard Animals for Livestock Protection: Existing and Potential Use in Australia". (Australia: Vertebrate Pest Research Unit, Orange Agriculture Institute, 2003) Found on the NSW Department of Primary Industries
www.dpi.nsw.gov.au

Rezendes, Paul. *Tracking and the Art of Seeing: How to Read Animal Tracks and Sign*. (HarperCollins, 1999)

Trout, John Jr. *Solving Coyote Problems: How to Outsmart North America's Most Persistent Predator*. (Lyons, 2001) America Sheep Industry Association's Sheep and Goat Research Journal: Special Issue: Predation. (Vol. 19, 2004)

Williams, Paul, ed. *Predator Control for Sustainable & Organic Livestock Production*. (Appropriate Technology Transfer for Rural Areas, 2002)
www.attra.ncat.org

DOG RESOURCES

Books and Articles

Buzady, Tibor. *Dogs of Hungary*. (Nora-Fonda-Herp, 2003)

Clothier, Suzanne. *Bones Would Rain from the Sky: Deepening Our Relationships with Dogs*. (Grand Central, 2005)

Coppinger, Raymond and Lorna. *Dogs: A Startling New Understanding of Canine Origin, Behavior, and Evolution*. (Scribner, 2001)

Dawydiak, Orysia and David Sims. *Livestock Protection Dogs: Selection, Care and Training*. (Alpine, 2004)

Derr, Mark. *A Dog's History of America: How Our Best Friend Explored, Conquered, and Settled a Continent*. (North Point, 2005)

Donaldson, Jean. *Culture Clash: A Revolutionary New Way to Understanding the Relationship Between Humans and Domestic Dogs*. (James & Kenneth, 1997)

Hancock, David. *The Heritage of the Dog*. (Nimrod, 1990)

Hubbard, Clifford L. B. *Working Dogs of the World*. (Sidgwick and Jackson, 1947)

"Livestock Guarding Dogs; Protecting Sheep from Predators." (USDA/APHIS, Information Bulletin Number 588, 1999) Found on the USDA National Agricultural Library
http://awic.nal.usda.gov

McConnell, Patricia B. *How to Be the Leader of the Pack . . . And Have Your Dog Love You for It*. (Dog's Best Friend, 1996)

DOG RESOURCES

Books and Articles (continued)

McConnell, Patricia B. *The Other End of the Leash.* (Ballantine, 2003)

Mehus-Roe, Kristin, ed. *The Original Dog Bible: The Definitive Source for All Things Dog.* (BowTie, 2005)

Morris, Desmond. *Dogs: The Ultimate Dictionary of Over 1,000 Dog Breeds.* (Trafalgar Square, 2002)

Oliff, Douglas, ed. *The Ultimate Book of Mastiff Breeds.* (Howell, 1999)

Pryor, Karen. *Don't Shoot the Dog.* (Ringpress, 2002)

Rigg, Robin. *Livestock Guarding Dogs: Their Current Use World Wide.* (Canid Specialist Group, 2001) Available on the Canid Specialist Group Web site: *www.canids.org*

Urbigkit, Cat. *Brave Dogs, Gentle Dogs: How They Guard Sheep.* (Boyd's Mills Press, 2005)

Magazines

(available at newsstands or by subscription)

AKC Gazette
American Kennel Club
800-533-7323
www.akc.org

Dog and Kennel
Pet Publishing, Inc.
336-292-4047
www.petpublishing.com/dogken

Dog Fancy
www.dogchannel.com/dfdc_portal.aspx

Dog World
www.dogchannel.com/dog-magazines/dogworld

Dogs in Canada
www.dogsincanada.com

Dogs USA
www.dogchannel.com/dog-magazines

Whole Dog Journal
800-424-7887
www.whole-dog-journal.com

Internet Resources

Black Dog Farm's Border Collie and Working Dog Links
www.geocities.com/black_dog_farm/BCLinks.html

Goats and Livestock Dogs
http://pets.groups.yahoo.com/group/goatslivestockdogs

Livestock Guardian Dogs
www.lgd.org

Working Livestock Guard Dogs Group
http://finance.groups.yahoo.com/group/workingLGDs/

Clubs and Associations

Most of these associations have information and breeder lists on their websites

American Kennel Club (AKC)
919-233-9767
www.akc.org

American Rare Breed Association (ARBA)
301-868-5718
www.arba.org

Canadian Kennel Club (CKC)
416-675-5511
www.ckc.ca

Federation Cynologique Internationale (FCI)
+32-71-59-12-38
www.fci.be

The Kennel Club (KC)
+44-0870-606-6750
www.the-kennel-club.org.uk

Livestock Guardian Dogs Association
www.lgd.org

United Kennel Club (UKC)
269-343-9020
www.ukcdogs.com

Breed Clubs

You can find a good breeder and a good dog through various sources, and you are encouraged to explore all of these options. Some breed clubs are affiliated with registries such as the AKC or UKC, while others have a specific focus. Inclusion on this list implies no endorsement of one club over another.

Akbash Dog
Akbash Dog Association of America (UKC)
www.turkishdogs.com/akbash

Akbash Dogs International
www.whitelands.com/akbash

Anatolian Shepherd Dog
Anatolian Shepherd Dog Club of America (AKC)
715-443-3509
www.asdca.org

Anatolian Shepherd Dogs International (UKC)
352-568-2557
www.anatoliandog.org

National Anatolian Shepherd Rescue Network
www.nasrn.com

Caucasian Mountain Dog/ Caucasian Ovcharka
American Association of Caucasian Ovtcharka Owners
406-889-5037
www.aacoo.net

Caucasian Mountain Dog Club of America
www.cmdca.com

Caucasian Ovcharka Club of America (UKC, FSS)
440-286-2374
www.cocaclub.us

Central Asian Shepherd Dog
Central Asian Shepherd Society of America (UKC, FSS)
716-751-6927
www.cassa.homestead.com

Estrela Mountain Dog
The Estrela Mountain Dog Association of America (FSS)
570-256-3976
www.emdaa.com

Estrela Mountain Dog USA
814-824-4906
http://estrelamountaindogusa.com

Great Pyrenees
Great Pyrenees Club of America (AKC)
847-658-7822
http://clubs.akc.org/gpca

Kangal Dog
Kangal Dog Club of America (UKC)
734-475-9633
www.kangalclub.com

Kangal Dog Preservation Trust
www.kangaldog.com

KyiApso
The Tibetan KyiApso Club
www.muddypaws.com/tkc.html

Komondor
Komondor Club of America (AKC)
607-565-9555
http://clubs.akc.org/kca

Middle Atlantic State Komondor Club
609-924-0199
www.maskc.org

Kuvasz
American Kuvasz Association
www.kuvasz.org

Kuvasz Club of America (AKC)
www.kuvasz.com

Kuvasz Fanciers of America
818-772-9364
http://members.aol.com/kfa4kuvasz

Kuvasz Club of Canada
905-276-4462
www.kuvaszclubofcanada.org

Maremma
The Maremma Sheepdog Club of America
715-364-2646
www.all-animals.com/maremma

Polish Tatra
Polish Tatra Sheepdog Club of America
540-752-1135
www.ptsca.com

Polish Tatra Sheepdog Club of Canada
www.polishsheepdog.ca

Pyrenean Mastiff
Pyrenean Mastiff Club of America (FCI)
661-724-0268
www.pyreneanmastiff.org

Rafeiro de Alentejo
Rafeiro do Alentejo USA Kennel
860-868-2936
ghenisz@sbcglobal.net

Slovak Cuvac
Slovac Cuvac Club of America
www.cuvac.us

Tibetan Mastiff
American Tibetan Mastiff Association
212-779-2715
www.tibetanmastiff.org

Tibetan Mastiff Club of America (TMCA)
www.tmcamerica.org

LLAMA RESOURCES

Books and Articles

Bennett, Marty McGee. *The Camelid Companion: Handling and Training Your Alpacas & Llamas.* (Raccoon Press, 2001)

Birutta, Gale. *Storey's Guide to Raising Llamas.* (Storey, 1997)

Cavalcanti, Sandra M. C., and Frederick F. Knowlton. "Evaluation of physical and behavioral traits of llamas associated with aggressiveness toward sheep-threatening canids" in Applied Animal Behaviour Science (1998 61: 43-158)
Found on the National Wildlife Research Center Web site: *www.aphis.usda.gov/ws/nwrc*

Hoffman, Clare and Ingrid Asmus. *Caring for Llamas and Alpacas: A Health Management Guide.* (Rocky Mountain Llama and Alpaca Association, 2003)

Magazines

Llama Banner
785-537-0320
www.llamabanner.com

Llama Life II
434-286-4494
www.llamalife.com

LamaLink.com
406-755-5473
www.lamalink.com

Clubs and Associations

Canadian Llama and Alpaca Association (CLAA)
800-717-5262
www.claacanada.com

International Llama Registry
406-755-3438
www.lamaregistry.com

Llama Association of North America
208-267-7398
www.llamainfo.org

Rocky Mountain Llama and Alpaca Association
www.rmla.com

Internet Resources

Llamapaedia
www.llamapaedia.com

LlamaWeb
www.llamaweb.com

Llama Life II
www.llamalife.com/archives/chuteplans.pdf
Design plans for building a llama chute

DONKEY RESOURCES

Books and Articles

Alberta Agriculture and Food. "Protecting Livestock with Guard Donkeys." (Alberta Agriculture and Food, 1994)
Found on the Alberta Agriculture and Food Web site: *www.agric.gov.ab.ca*

Gross, Bonnie R. *Caring for Your Miniature Donkey.* (Miniature Donkey Talk, 1998)

Hutchins, Betsy and Paul. *The Definitive Donkey: A Textbook on the Modern Ass.* (American Donkey & Mule Society, 1999)

Tapscott, Brian. "Guidelines for Using Donkeys as Guard Animals with Sheep." (Ontario Ministry of Agriculture, Food, and Rural Affairs, 1997) Found on the Ontario Ministry of Agriculture, Food, and Rural Affairs Web site: *www.omafra.gov.on.ca*

Texas Department of Agriculture. "Using Donkeys to Guard Sheep and Goats." (Texas Department of Agriculture, 2006) Found on the Texas Department of Agriculture Web site: *www.agr.state.tx.us*

Magazines

The Brayer
American Donkey and Mule Society
972-219-0781
www.lovelongears.com

Clubs and Associations

American Council of Spotted Asses
636-828-5955
www.spottedass.com

American Donkey and Mule Society
972-219-0781
www.lovelongears.com

Canadian Donkey & Mule Association
403-742-1144
www.donkeyandmule.com

PHOTO CREDITS

© John Anderson/iStockphoto: 189

© Nicholay Atanassov: 110 bottom

© Holly and Henry Ballester: 87

© Feride and Hasan Cansever: 109 top left

© Audrey Chalfen: 70

© John Daniels/ardea.com: 112 top left, 113 top left

© Jeanne DePalma: 112 top right

© Mark Duffy/Painet, Inc.: 115

© Kathy Gerlach/Gerlach Ranch Anatolians: 39

© Bruce J. Gilbert/Painet, Inc.: 116

© Nicky Gordon/iStockphoto: 112 bottom left

© Karin Graefe/De La Tierra Alta Kennels: 11 bottom, 113 top right

© Onur Kanli: 109 top right

© Jean Michel Labat/ardea.com: 11 top left

© Craig Lovell/Painet, Inc.: 109 bottom

© Luke Peters/Alamy: 114 top left

© Linda Raeber: 110 top

© Sherri Regalbuto: 112 bottom right

© Stuart Dudley Richens: 75

© Lynn Siler Photography/Alamy: 114 top right

© Diane Spisak/Sheepfields, Akbash Dogs, Babydoll Southdowns & Polish Bantams: 45, 113 bottom

© Sabine Stuewer: 11 top right

United States Department of Agriculture: 47, 91

© Cat Urbigkit: 103–108, 114 bottom, 117, 188 top

© Elizabeth von Buchwaldt: 41

© Andreas Weissen: 118 bottom

INDEX

Page numbers in *italic* indicate illustrations; those in **bold** indicate tables.

A

abscesses
    donkey's foot, 207, 209
    llama's molar, 181
additional dogs, introducing, 72
adopting
    donkeys or burros, 194, 197
    rescue dogs, 55–56, 70–71
    rescue llamas, 167
adult dogs as mentors, 51, 57, 59, 63, 72, 92, 107, *107,* 160
adult vs. young
    dogs, 50, *63,* 63–64, 70–71, *71*
    donkeys, 191–92, 198
    llamas, 167, 168, 185
Afghanistan, *102,* 156, *156*
African ass *(Equus asinus),* 187
afterbirth, dogs eating, 62
aggression
    dogs, 74, 77–78, *78,* 79
    donkeys, 209, 210
    llamas, 166, 168–69, 184–85
Akbash Dog, 45, *45,* 98, *98, 102,* 113, *113, 143,* 143–44, 146, 149, 216
alert behavior, llamas, 165, 166, 167
alpacas, 79, 163, *163,* 164, 165
American Donkey and Mule Society, 195, 204
American Kennel Club (AKC), 43, 99, **99,** 119, 126, 132, 134, 153, 155, 216
American Rare Breed Association (ARBA), **99,** 144, 216
American Standard (Mammoth Jack), *195,* 195–96, 197
America Sheep Industry Association, 2
ammunition (nonlethal), 37

Anatolian Shepherd Dog, 39, *39,* 45, *102,* 110, *110, 149,* 149–51, 216
anatomy of
    dog, *85*
    donkey, *201*
    llama, *177*
Andelt, William, 43
Anderson, Dean M., 37
Andrean fox, 164
animal management, 23, **25,** 25–27
anterior cruciate ligament (ACL) injuries, dogs, 93
anthrax caution, 5
APHIS (U.S. Animal and Plant Health Inspection Service) Wildlife Services, 35, 213
approaching a donkey, 193
Armenia, *102,* 151, *151*
Armenian Shepherd (Gampr), *102,* 151, *151*
arthritis, dogs, 93
ascarids, 205
Askaray Dogs, 149
Athens Canine Society, 139
attack by dog, what to do, 8
attacks by donkeys, 191
auditory deterrents, 35–36
aversive predator control methods, 23, **25,** 36–37

B

barbed wire caution, llamas, 179
barking (excessive), dogs, 79–80
Bearded Mastiff (KyiApso), 4, *102, 157,* 157–58, 217
bears, 2, 3, **3,** 5, 14–17, *15,* 27
behavioral characteristics

dogs, 39, *45,* 45–48, *47–48*
donkeys, 187, 188–91, *189–90*
llamas, 162, 164–67
bells as auditory deterrent, 35
Belyaev, Dmitry K., 47
binocular vision, donkeys, 189
biological deterrents, 36
birds (large), 18–19
birthing season
    dogs and, 61–62
    donkeys and, 210
    llamas and, 185
black bears *(Ursus americanus),* 14–15, *15,* 16–17
black vultures, 19
BLM (U.S. Bureau of Land Management), 117, 188, 194, 197
bloat, 92, 178
blood work, llamas, 179
bobcats *(Felis rufus),* 3, **3,** 5, 13–14, *14*
bonding. *See* socialization and bonding
bone issues, dogs, 92
bones, feeding to dogs, 86
Border collies, 43, 48
bots, and donkeys, 205, 206
braying (excessive) donkeys, 210–11
breeders
    dogs, 4, 48, 52, 53–54, 74
    donkeys, 191, 194–95
    llamas, 167, 168
breeding your dog, 94–95
breed registries (dogs), **99,** 216
breeds
    dogs, 96–159
    donkeys, *195,* 195–97
    llamas (fleece), 163, 170–73
bringing your puppy home, 59, *59*

brown bears *(Ursus arctos)*, 15, *15*, 16–17
Bukovina Shepherd (Ciob nesc Românesc de Bucovina), *101*, *141*, 141–42
Bulgaria, *101*, 139–41, *140*
Bulgarian Biodiversity Preservation Society Semperviva, 139, 140
Bulgarian Shepherd Dog (Karakachan), *101*, 110, *110*, 139–41, *140*, 142, 160
burros, *117–18*, **187**, 188
butchering caution, dogs, 61

C

Camelidae family, 162. *See also* llamas (livestock guardians)
Canadian Horse and Mule Association, 195
Canadian Kennel Club (CKC), **99,** 216
Canadian lynx *(Felis lynx)*, 3, 5, 13–14, *14*
Canidae family, 40. *See also* dogs (livestock guardians)
Canine Eye Registration Foundation (CERF), **99,** 214
canine inherited demyelinative neuropathy (CIDN), 159
Cannon-Lelli, Brenda (Beechtree Farm, Coopersville, MI), 76
Cão da Serra da Estrela (Estrela Mountain Dog), *101*, 112, *112*, *119*, 119–20, 216
Cão de Castro Laboreiro (Portuguese Cattle Dog, Castro Laboreiro), *101*, *120*, 120–21
Cão de Gado Transmontano (Rafeiro Montano, Rafeiro Transmontano), *101*, 122, *122*
caring for your livestock guardians. *See also* selecting and training dogs, 84–95

donkeys, 200–209
llamas, 176–84
carnivore conservation groups, 44
Carpathian Large Carnivore Project, 141–42
Carpathian Shepherd (Ciob nesc Carpatin), *101*, 142, *142*
Castro Laboreiro (Cão de Castro Laboreiro), *101*, *120*, 120–21
catamount. *See* cougar
catching issues, donkeys, 211
catching tip, llamas, 172
catch pen or chute, llamas, 169, 173–74, 175, *175*, 181
cats (feral and domestic), 18, 21
Catskill Game Farm, 163
cattle and dogs, 79
Caucasian Mountain Dog/Caucasian Ovcharka (Caucasian Shepherd, Kavkazskaïa Ovtcharka), 54, *102*, 112, *112*, 137, 151, *153*, 153–54, 216
Cavalcanti, M. C., 166
Ccara (Kcara) llamas, 163, 171, 182
Celebi, Evliya, 143
Central Asia, *102*, *154*, 154–55
Central Asian Shepherd Dog (Central Asian Ovcharka, Sredneasiatskaia Ovtcharka), *102*, 112, *112*, 153, *154*, 154–55, 216
chaku llama, 163
chasing livestock by dogs, 80, 81, *81*
checking livestock, 3–5, **25,** 26
chemical deterrents, 36
chewing (inappropriate), dogs, 80, 82
Chien de Montagne des Pyrénées (Great Pyrenees), 43, *43*, 45, 46, 47, *47*, 76, 98, *101*, 106, *106*, 111, *111*, 125–27, *126*, 134, 194, 216
children and dogs, 75, *75*, 77
choke, in llamas, 183

choke (slip) collars, 65–66, *66*, 67
Ciob nesc Carpatin (Carpathian Shepherd), *101*, 142, *142*
Ciob nesc Românesc de Bucovina (Bukovina Shepherd), *101*, *141*, 141–42
Ciob nesc Românesc Mioritic (Mioritic Sheepdog), *101*, 142, 143, *143*, 152
clubs (dog), 216–17
coban kopegi, 149, 150
coccidia, 180
Coggins test, 194, 203
colic, 183, 202, 206–7
collars, dogs, 65–67, *66*, 69
Colorado State University, 43
combination nonlethal predator control, 35, 37
communication, llamas, 164, 165
companion or house guard dogs, 39, 40, 48
condors, 164
Coppinger, Ray, 120, 121, 128, 150
Coppinger, Ray and Lorna, 43
corrections, dogs, 61, 62, 65, 70
costs of livestock guardians
dogs, 45, 50, 54, 55
donkeys, 190, 194–95
llamas, 167
cougars *(Puma concolor)*, 2, 3, **3,** 5, 12–13, *13*, 27
coydogs, 11
coyotes *(Canis latrans)*, 2, 3, **3,** 5–7, *6*, *8*, 11, 23
coywolves, 11
crate training, 60
cria, **163**
Croatia and Bosnia Herzegovina, *101*, *136*, 136–37
crossbred dogs, 54–55
crossbred llamas, 173
crossbreeding (wolves, coyotes, jackals, dogs), 11
Curaca llamas, 171

Cuvac (Slovak Cuvac), *101*, 131, *131*, 217

D

dangle sticks, 81, *81*
dead livestock, 3–5, 25, 82
death of dogs, causes of, **67**, 68
deer problems and dogs, 48
Demodex mite, 91
dental problems, llamas, 181
dermatitis, dogs, 90–91
deskunking solution, 94
deworming
    donkeys, 205–6
    llamas, 180
Dhokhi Apso (KyiApso), 4, *102*, *157*, 157–58, 217
dingoes, 164
disruptive predator control, 23, **25**, 33–36, *35–36*
distemper caution, 5
diversionary feeding, 37
dog attack, what to do, 8
dog fighting, 99
dogs (free-roaming and feral), 2, 3, **3**, 7–9, *8*, 11
dogs (livestock guardians), **25**, 38–48
    anatomy of, *85*
    behavioral characteristics, 39, *45*, 45–48, *47–48*
    breeds of, 96–159
    caring for, 84–95
    costs, 45, 50, 54, 55
    domestication of, 40, 41, 45, 46–47
    donkeys and, 198, 210
    feeding, 85–87, *87*, 92
    fencing, *68*, 68–70, *70*, 74, 75
    health care, 88–94, **89**, *91*
    life span of, 91, 99
    llamas and, 79, 162, 167, 170, 185
    origins and history, 40–45, *41–45*
    other dogs and, 59–60
    pros and cons, **40**, 45
    research on, 43
    resources for, 215–17
    selecting and training, 49–83
    socialization and bonding, 45–47, *47*, 48, 54, 57, 58, 59–60, 73, 74, 77
    solving problem situations, 77–83
    working livestock guard, 39, 40
Do-Khyi (Tibetan Mastiff), *102*, 109, *109*, 157, *158*, 158–59, 217
domestication of
    dogs, 40, 41, 45, 46–47
    donkeys, 187
    llamas, 163
donkeys (livestock guardians), **25**, 186–91
    anatomy of, *201*
    behavioral characteristics, 187, 188–91, *189–90*
    caring for, 200–209
    costs, 190, 194–95
    dogs and, 198, 210
    domestication of, 187
    feeding, 199, 201–2, 207, 209, 210, 211
    fencing, 202
    health care, 202–9, *204*, *208*
    horses vs., 195, 202, 203, 205, 206
    life span of, 187
    origins and history, 187–88
    pros and cons, 191, **192**
    research on, 190
    resources for, 218
    selecting and training, 191–99
    socialization and bonding, 188, 189, *189*, 191, 198
    solving problem situations, 209–11
    working livestock guard, 190, *190*
drags, 81
drift fences, 27, **28**
driving your donkey, 211

E

eagles, 3, **3**, 18–19
ear care, dogs, 90
Eastern Europe, *102*, *152–53*, 152–54
ectropion, 92
elbow dysplasia, dogs, 94
electric fencing, 27, **28**, 29–30, *30*, 31, 32
electric net fencing, 30
electronic training collars, 36–37
El Mastin del Pirineo (Pyrenean Mastiff), *101*, 111, *111*, *124*, 124–25, 217
Endangered Species Act, 3
entropion, 92
equine infectious anemia (EIA), 194, 203
equine protozoal myeloencephalitis (EPM), 5
Estrela Mountain Dog (Cão da Serra da Estrela), *101*, 112, *112*, *119*, 119–20, 216
European Radiology Association, 125
exclusion fences, 27, **28**
existing fence, upgrading, **28**, 31–32, *32*
eye care, dogs, 90, 92

F

failure to work
    dogs, **67**, 68, 83
    donkeys, 211
    llamas, 185
Faktor, Cherie (Willow Ridge Farm, Lucas, IA), 34

family companion dogs, 74–77, *75*
farm and family guardian dogs, 39, 48, 50, *72,* 72–74
fecal egg count, 205
Fédération Cynologique Internationale (FCI), 97, **99,** 120, 123, 124, 130, 131, 133, 136, 137, 139, 142, 143, 147, 151, 152, 153, 155, 159, 216
Federation of Dog Breeding and Cynology, 147
feeding
    dogs, 85–87, *87,* 92
    donkeys, 199, 201–2, 207, 209, 210, 211
    llamas, 177–78, *178,* 182, 183
females vs. males
    dogs, 51
    donkeys, 192–93, 210
    llamas, 167, 168, 177, 181
fenbendazole, 180, 205
fencing, x, 2, 23, **25,** 27–33. *See also* nonlethal predator control
    dogs and, *68,* 68–70, *70,* 74, 75
    donkeys and, 202
    drift fences, 27, **28**
    electric fencing, 27, **28,** 29–30, *30,* 31, 32
    exclusion fences, 27, **28**
    gates, 31, 32–33, *32–33*
    guidelines, 27–28
    installing a new fence, **28,** 30–31, *31*
    llamas and, 179
    upgrading existing fence, **28,** 31–32, *32*
    wire mesh (mesh) fencing, 27, 28–29, *28–29,* 30–31
feral and free-roaming dogs, 2, 3, **3,** 7–9, *8,* 11
feral cats, 18, 21
feral hogs *(Sus scrofa),* 3, 5, *17,* 17–18
fertilizer, llama manure as, 185

Field, C. Andy, 190
fighting teeth, llamas, 168, 181
fishers *(Martes pennanti),* 18, 21
fixed noseband halters, 174
fladry, visual deterrent, 34–35, *35*
fleas, 89, **89**
fleece of llama, using, 182
fleece types, llamas, 163, 170–73
flight response, donkeys, 189, 191
floating (filing) teeth, 181, 204
foal (donkey), starting with, 191–92, 193, 198
Food and Agricultural Organization of the United Nations (FAO), 97, 156
foot care
    dogs, 90
    donkeys, 193, 203–4, *204,* 207–9, *208*
    llamas, 169, 172, 175, 177, *180,* 180–81
founder (laminitis), 202, 203, 207–8, *208*
fox domestication experiment, 47
foxes, 3, **3,** 18, 19–20
France, *101,* 125–27, *126*
Franklin, William L., 166
free-roaming and feral dogs, 2, 3, **3,** 7–9, *8,* 11

G

Gampr (Armenian Shepherd), *102,* 151, *151*
gates, 31, 32–33, *32–33*
geldings
    donkeys, 190, 192
    llamas, **163,** 167, 168, 177
genetic issues, dogs, 53, 54, 99
genotype, defined, **97**
goat losses, 2–3, **3**
gray wolves *(Canis lupus), 9,* 9–10, 40
grazing, nonlethal predator control, 26

Great Pyrenees (Pyrs, Pyrenean Mountain Dog, Chien de Montagne des Pyrénées), 43, *43,* 45, 46, 47, *47,* 76, 98, *101,* 106, *106,* 111, *111,* 125–27, *126,* 134, 194, 216
Greece, *101, 138,* 138–39
Greek Sheepdog (Hellenikos Poimenikos), *101, 138,* 138–39, 142
Green, Jeff, 43
grizzly bears *(Ursus arctos horribilus),* 15, *15,* 16–17
grooming
    dogs, 91, *91*
    donkeys, 206
    llamas, 164, 181–82
Grupo Lobo, 119, 121, 213
guanacos, 163, 164

H

hairworms, 205
halters and halter training
    donkeys, 199
    llamas, 169, *174,* 174–75
Hampshire College, 43, 120, 121, 128
hand (measurement), **187**
hawks, 18–19
head halters, dogs, 66, *66,* 67
health care
    dogs, 88–94, **89,** *91*
    donkeys, 202–9, *204, 208*
    llamas, 179–84, *180–81, 183*
health certificates, 168, 193, 194
health registries, dog, **99,** 214
health threats from wild animals, 5, 20, 21, 48
hearing sense, donkeys, 189
heartworms, **89,** 89–90
heat stress, llamas, 182, 183–84
Hellenikos Poimenikos (Greek Sheepdog), *101, 138,* 138–39, 142

herding (active), **25,** 26
herding vs. livestock guardian dogs, 43, 55, 78
hinny, **187**
hip dysplasia, dogs, 53, 54, 93–94, 99, **99**
hogs, feral *(Sus scrofa)*, 3, 5, *17,* 17–18
hookworms, 5, 89, **89**
horses
   dogs and, 62
   donkeys vs., 195, 202, 203, 205, 206
   llamas and, 173
   wild and feral, 188
Hrvatski Ovcar (Tornjak), *101, 136,* 136–37, 142
huacaya alpaca, 163
humans, aggression from
   dogs, 77–78
   donkeys, 209
   llamas, 168–69, 184
Hungary, *101, 132–33,* 132–34
hunting habits of
   bears, 16–17
   coyotes, 7
   dogs, 7–9
   feral hogs, 18
   foxes, 20
   large birds, 19
   lynx and bobcats, 14
   mountain lions, 12–13
   raptors, 19
   wolves, 10
husbandry and animal management, 23, **25,** 25–27
hybridization (wolves, coyotes, jackals, dogs), 11
hypothermia, llamas, 184

I

identification, dogs, 65
illness (signs of)
   donkeys, 206, 208, *208,* 209
   llamas, 182, 183, *183,* 184
Illyrian Sheepdog (Sarplaninac), 45, *101,* 135, *137,* 137–38, 142, 153
Il Pastore Maremmano Abruzzese (Maremma Sheepdog), 45, *101,* 112, *112,* 127–29, *128,* 217
inherited guardian traits, dogs, 48, *48*
injured livestock and dogs, 82
installing new fence, **28,** 30–31, *31*
Institute of Cytology and Genetics (Siberia), 47
insurance, 77
integrated plan of protection, 23, 24, *24,* 25
International Llama Registry, 163
introducing
   additional dogs, 72
   adult dog to livestock, 70–71, *71*
   donkey to livestock, 197, *198,* 198–99
   llama to livestock, 173
   new livestock to dogs, 72
   puppy to livestock, 60–64, *61–63,* 103–6, *103–7,* 107
invisible (radio) fencing, *68,* 68–70
Iowa State University, 165, 166, 185
isolation method, donkey training, 198–99
Istrian Sheepdog (Karst Shepherd), *101,* 135, *135,* 137
Italy, *101,* 127–29, *128*
ivermectin, **89,** 90, 180, 205, 206

J

jacks, **187,** 188, 192–93, 210
jennet, jenny, jennet filly, **187**
jennies, 188, 189, *189,* 190, 192–93, 210
joint issues, dogs, 92
jump feeder, 87, *87*

K

Kangal Dog (Kangal Kopegi), 41, *41,* 57, 75, *75, 102,* 109, *109,* 143, 144, *146,* 146–48, 149, 216
Karakachan (Bulgarian Shepherd Dog), *101,* 110, *110,* 139–41, *140,* 142, 160
Kars Dog, *102,* 148–49, *149*
Karst Shepherd (Kraski Ovcar, Krasky Ovcar, Krasevec, Istrian Sheepdog), *101,* 135, *135,* 137
Kavkazskaïa Ovtcharka (Caucasian Mountain Dog), 54, *102,* 112, *112,* 137, 151, *153,* 153–54, 216
Kcara (Ccara) llamas, 163, 171, 182
Kennel Club, UK (KC), **99,** 216
kenneling your dog, 82
Knowlton, F. F., 166
Komondor, 43, 45, 91, *91, 101,* 113, *113, 132,* 132–33, 134, 152, 217
Koochee (Sage Koochee), *102,* 156, *156*
Kraski Ovcar, Krasky Ovcar, Krasevec (Karst Shepherd), *101,* 135, *135,* 137
kush, **163,** 168, 179
Kuvasz, 43, 45, *101,* 111, *111,* 131, *133,* 133–34, 217
KyiApso (Dhokhi Apso, Lhasa Apso, Bearded Mastiff), 4, *102, 157,* 157–58, 217

L

laminitis (founder), 202, 203, 207–8, *208*
landrace breeds, **97,** 98, 99
Lane, Dan and Paula (Bountiful Farm, LeFlore County, OK), 46

Lanuda llamas, 171
Large Carnivore Initiative for Europe, 44, 213
large predators, 5–18. *See also* predation, problem of
Lavooi, Cheryl (Wooly Manor, Republic, WA), 172
leadership (establishing your), dogs, 64–65, 78
leadership of flock, llamas, 165–66
leading llamas, 175
leash training, dogs, 60, 61, 64, *64*
lethal predator control, 23, 24, 42, 43
Lhasa Apso (KyiApso), 4, *102*, *157*, 157–58, 217
lice, donkeys, 205
life span of
    dogs, 91, 99
    donkeys, 187
    llamas, 167
lighting as visual deterrent, 33–34
Limpus, Mary and Larry (Limpus Farms, Amsterdam, MO), 194
liver flukes *(Fasciola)*, 180
livestock, aggression toward dogs, 79
livestock, aggression from
    donkeys, 210
    llamas, 184–85
livestock, introducing
    adult dog to, 70–71, *71*
    donkey to, 197, *198*, 198–99
    llama to, 173
    puppy to, 60–64, *61–63*
Livestock Guard Dog Project (Hampshire College), 43, 120, 121
livestock guardians. *See* dogs (livestock guardians); donkeys (livestock guardians); llamas (livestock guardians); predation, problem of; predator control methods

livestock protection collars, 23, 36, *36*
llamas (livestock guardians), **25,** 161–67
    anatomy of, *177*
    behavioral characteristics, 162, 164–67
    caring for, 176–84
    costs, 167
    dogs and, 79, 162, 167, 170, 185
    domestication of, 163
    feeding, 177–78, *178*, 182, 183
    fencing, 179
    fleece types, 163, 170–73
    health care, 179–84, *180–81*, *183*
    life span of, 167
    origins and history, 162–64
    predators (natural) of, **164**
    pros and cons, 162, *162*, **166**
    research on, 164–65
    resources for, 217
    selecting and training, 167–75
    socialization and bonding, 162, *162*, 164, 168, 169, *169–70*, 170, 173
    solving problem situations, 184–85
    working livestock guard, 164–67
losses from predators, 2–3, **3**
loss of dog, causes of, **67,** 68
lungworms *(Dictyocaulus arnfieldi)*, donkeys, 106, 205
Lyme disease, 89
lynx *(Felis lynx)*, 3, 5, 13–14, *14*

M

M-44s (baited cyanide injectors), 23
Macedonia, *101*, *137*, 137–38
maiden llama, **163**
males vs. females
    dogs, 51
    donkeys, 192–93, 210

    llamas, 167, 168, 177, 181
Mammoth Jack (American Standard), *195*, 195–96, 197
mange, dogs, 5, **89,** 91
mange mites, donkeys, 205
manure (llama) as fertilizer, 185
Maremma Sheepdog (Il Pastore Maremmano Abruzzese), 45, *101*, 112, *112*, 127–29, *128*, 217
mastiffs *(molosser)*, 41, 158
Mastín Español Club of America, 123
Mastín Español (Spanish Mastiff), 54, *101*, 113, *113*, 121, *123*, 123–24, 125
mastino (Pyrenean Mastiff), *101*, 111, *111*, *124*, 124–25, 217
matron llama, **163**
meningeal worms *(Parelaphostrongylus)*, 180
mesh fencing, 27, 28–29, *28–29*, 30–31
Mibemycin oxime (Interceptor), **89**
Mibemycin oximeleufenuron (Sentinel), **89**
minerals and salt, 178, 202
Miniature Mediterranean donkey, 188, 196
minks, 18, 20
Mioritic Sheepdog (Ciob nesc Românesc Mioritic), *101*, 142, 143, *143*, 152
monocular vision, donkeys, 189
Moscow Watchdog, 153
mountain lions *(Puma concolor)*, 2, 3, **3,** 5, 12–13, *13*, 27
moxidectin, 180, 205
mules, **187,** 188
multispecies grazing, 37

N

National Agricultural Statistics Service (NASS), 2, **3, 25,** 213
National Wildlife Research Center, 48, 213
necked strongyles *(Nematodirus)*, 180
neoteny, 47
neutering dogs, 51, 70
new fence, installing, **28,** 30–31, *31*
new livestock and dogs, 72
New Mexico State University, 37
night confinement, x, 2, 26
nodular worms *(Oesphagostomum)*, 180
nonlethal predator control, 23–37, **25**. *See also* dogs (livestock guardians); donkeys (livestock guardians); fencing; llamas (livestock guardians); predator control methods
Nubian wild ass *(Equus asinus africanus)*, 187. *See also* donkeys (livestock guardians)

O

older dogs, caring for, 91–92
open female llama, **163**
opossums *(Didelphis virginiana)*, 5, 21
Orthopedic Foundation for Animals (OFA), 94, **99,** 146, 214
osteoarthritis, dogs, 93
osteochondrosis (OCD), dogs, 93, 94
osteosarcoma, dogs, 94
Ovcarski Pas (Sarplaninac), 45, *101,* 135, *137,* 137–38, 142, 153
ovcharka group, 153
Ovcharka (South Russian Ovcharka), *102,* 152, *152*
owls, 18–19
oxibendazole, 205

P

"pack" of guard dogs, 10
pairs (dogs), training, 60
panosteitis, dogs, 93
parasites
  dogs, **89,** 89–90
  donkeys, 202, 205–6
  llamas, 180
  wild animals and, 5
Parque Natural da Montesinho, 122
parvo caution, 5
Patterson llama herd (OR), 163
pedigree, defined, **97**
PennHIP (University of Pennsylvania Hip Improvement Program), 53, 94, **99,** 214
pens for puppies, 59, *59*
pets (other) and dogs, 76–77, 78
phenotype, defined, **97**
physical corrections, dogs, 61, 62
pinch (training) collars, 66, *66,* 67
poison bait, 23
poison caution
  donkeys, 202
  llamas, 177–78
Poitou, 195, 196–97
Poland, *101, 129,* 129–30
Polish Tatra Sheepdog (Polski Owczarek Podhalanski, Tatra), *101, 129,* 129–30, 131, 217
porcupine quills, 95
porthole (dog door), 68, 87, *87*
Portugal, *101,* 119–22, *119–22*
Portuguese Cattle Dog (Cão de Castro Laboreiro), *101, 120,* 120–21
Portuguese Sheepdog (Cão Serra de Aires), 120
Portuguese Watchdog (Rafeiro do Alentejo), *101, 121,* 121–22, 217
poultry, 26–27, *62,* 62–63
Poyner, Robyn and Jim (Poyner Goat Company, Monett, MO), 145

prairie wolf *(Canis latrans),* 2, 3, **3,** 5–7, *6, 8,* 11, 23
predation, problem of, 1–21. *See also* predator control methods
  checking on livestock, importance of, 3–5, **25,** 26
  dead animals, determining cause of death, 3–5
  health threats from wild animals, 5, 20, 21, 48
  large predators, 5–18
  losses from predators, 2–3, **3**
  night confinement, x, 2
  predation studies, 2
  resources for, 213, 214
  signs of predation, 3–5, 82
  small predators, 18–21
Predator Conservation Alliance, 25, 213
predator control methods, 22–37. *See also* dogs (livestock guardians); donkeys (livestock guardians); fencing; llamas (livestock guardians); predation, problem of
  lethal predator control, 23, 24, 42, 43
  nonlethal predator control, 23–37, **25**
predator-friendly marketing, 25, 214
predatory behavior of dogs, 47–48
prepurchase exams, 168, 193
problem situations. *See* solving problem situations
prong (training) collars, 66, *66,* 67
pros and cons of
  dogs, **40,** 45
  donkeys, 191, **192**
  llamas, 162, *162,* **166**
protection training, dogs, 48, 73
pseudorabies caution, 5
Puli, 91
puma. *See* cougar

puppies. *See also* selecting and training, dogs
   adult dog as mentor, 51, 57, 59, 63, 72, 92, 107, *107,* 160
   bringing your puppy home, 59, *59*
   feeding, 86
   introducing to livestock, 60–64, *61–63,* 103–6, *103–7,* 107
   purchasing, 50, *51,* 51–54, *53,* 76
   raising and training, *58–59,* 58–60
   selecting as livestock guardians, 48, *48*
pyrantel, 180, 205
Pyrenean Mastiff (El Mastin del Pirineo, *mastino*), *101,* 111, *111, 124,* 124–25, 217
Pyrenean Mountain Dog, Pyrs (Great Pyrenees), 43, *43,* 45, 46, 47, *47,* 76, 98, *101,* 106, *106,* 111, *111,* 125–27, *126,* 134, 194, 216

Q

quarantine of dog, 70–71

R

rabies
   vaccination, 89, 180, 202
   wild animals and, 5, 20, 21, 160
raccoons *(Procyon lotor),* 5, 18, 20
radio as auditory deterrent, 35
radio (invisible) fencing, *68,* 68–70
Rafeiro do Alentejo (Portuguese Watchdog), *101, 121,* 121–22, 217
Rafeiro Montano, Rafeiro Transmontano (Cão de Gado Transmontano), *101,* 122, *122*
rain rot, 179, 209

raising and training your puppy, *58–59,* 58–60
raptors, 3, **3,** 18–19
rats, 18, 21
raw food, dogs, 86
recall, dogs, 64
record keeping, 26
red foxes *(Vulpes vulpes),* 19–20
red wolves *(Canis rufus),* 10–12, *12*
rescue dogs, 55–56, 70–71
rescue llamas, 167
research on livestock guardians
   dogs, 43
   donkeys, 190
   llamas, 164–65
respiration, llamas, 182, 184
respiratory disease, donkeys, 209
restraining llamas, 175, *175*
Richens, Stuart and Bob (Banks Mountain Farm, Hendersonville, NC), 57
riding your donkey, 211
Romania, *101,* 141–43, *141–43*
rotational grazing, nonlethal predator control, 26
roundworms, 89, **89**
Russian Kynological Federation (RKF), 153

S

Sage Koochee (Sage Mazandarani, Sage Rama, Koochee, Djence Palangi, Djence Sheri), *102,* 156, *156*
Sanders, Peter and Joan (Houseman Acres, 100 Mile House, BC), 16
Sarcoptes mite, 91
Sarplaninac (Illyrian Sheepdog, Ovcarski Pas, Yugoslavian Mountain Dog), 45, *101,* 135, *137,* 137–38, 142, 153
Savolainen, Peter, 143
scabies, dogs, 91

scent marking (artificial) as biological deterrent, 36
screening, nonlethal predator control, 26–27
Sedechev, Sider and Atilia, 140
Selamectin (Revolution), **89**
selecting and training
   dogs, 49–83
   donkeys, 191–99
   llamas, 167–75
self-feeding, dogs, 86–87
shearing llamas, 172, 173, 175, 182, 183
sheepdogs, 41
sheep losses, 2–3, **3**
shelter
   dogs, 88, *88*
   donkeys, 202, 203, *203*
   llamas, 178–79, 183
shepherds and dogs, 44, *44,* 51, 58, 63
shock collars, 79–80, 81
sight sense, donkeys, 189
signs of predation, 3–5, 82
single vs. multiple llamas, 166, 167, 170
sire (stud), llama, **163**
size and aggressiveness, llamas, 166
size of donkey, 191, 195, *195*
skin problems, dogs, 90–91
skunks, 5, 20–21
skunk spray, 94
Slovak Cuvac (Slovensky Cuvac, Cuvac), *101,* 131, *131,* 217
Slovakia, *101,* 131, *131*
Slovenia, *101,* 135, *135*
small predators, 18–21. *See also* predation, problem of
smell sense, donkeys, 189
socialization and bonding
   dogs, 45–47, *47,* 48, 54, 57, 58, 59–60, 73, 74, 77
   donkeys, 188, 189, *189,* 191, 198

socialization and bonding *(continued)*
  llamas, 162, *162*, 164, 168, 169, *169–70*, 170, 173
solving problem situations. *See also* selecting and training
  dogs, 77–83
  donkeys, 209–11
  llamas, 184–85
sound devices as auditory deterrent, 35–36
South Russian Ovcharka (South Russian Shepherd Dog, Ovcharka, Youzhak), *102*, 152, *152*
Spain, *101*, *123–24*, 123–25
Spanish colonization and llamas, 163
Spanish Jack, 196
Spanish Mastiff (Mastin Español), 54, *101*, 113, *113*, 121, *123*, 123–24, 125
spike collars, 41, *41*, 109, *109*
spitting, llamas, 164, 165
Sponenberg, Torsten and Phil (Beechkeld Farm, Blacksburg, VA), 160
Sredneasiatskaia Ovtcharka (Central Asian Shepherd Dog), *102*, 112, *112*, 153, *154*, 154–55, 216
Standard donkey, 195, *195*
standardized dog breeds, developing, *98*, 98–99, **99**
Steffel, Judy (Barking Dog Goat Farm, Houghton, MI), 4
stomach worms *(Haemonchus, Ostertagia, Trichostrongylus)*, 180, 205
stronglyes (large and small), 205
stubbornness, donkeys, 189–90
submissive behavior of dogs, 53, *53*, 58, *58*
supplements, dogs, 93
Suri llamas, 163, 172
sustainable agriculture, 24–25

swine brucellosis caution, 5

T

Taenia hydatigena, ovis, 89
Tampuli llamas, 171
tapeworms, 89, **89,** 180, 205
Tatra (Polish Tatra Sheepdog), *101, 129*, 129–30, 131, 217
teeth, care of
  dogs, 90
  donkeys, 204
  llamas, 181, *181*
temperature, llamas, 182, 184
threadworms, 180, 205
thrush, donkeys, 203, 208
thunderstorms and dogs, 82–83
Tibet, *102, 157–58*, 157–59
Tibetan Mastiff (Do-Khyi, Bhotia, Tsang-Khyi), *102*, 109, *109*, 157, *158*, 158–59, 217
ticks *(Dermacentor variabilis)*, 89, **89**
Tornjak (Hrvatski Ovcar), *101, 136*, 136–37, 142
tracks of
  bears, 15, *15*
  dogs and coyotes, 8, *8*
  feral hogs, 18
  mountain lions, 13
  wolves, 12
training. *See* selecting and training
training collars, 66, *66*, 67
transporting animals across state lines, 214
treats caution, donkeys, 209, 210
trichinosis caution, 5
trimming. *See* foot care.
trip wire (electric), 32, 33, *33*
Tsang-Khyi (Tibetan Mastiff), *102*, 109, *109*, 157, *158*, 158–59, 217
tuberculosis caution, 5, 48
tularemia caution, 5
turbo fladry as visual deterrent, 35

Turkey, *102, 143*, 143–44, *146*, 146–51, *149*
two-way adjustable halters, 174
tying
  dogs, 64, *64*
  llamas, 175
type, dogs, **97,** 98, *98*, 99

U

ultrasonic devices as auditory deterrent, 36
United Kennel Club (UKC), 99, **99,** 119, 137, 144, 147, 153, 155
University of Pennsylvania Hip Improvement Program (PennHIP), 53, 94, **99,** 214
upgrading existing fence, **28,** 31–32, *32*
U.S. Animal and Plant Health Inspection Service (APHIS) Wildlife Services, 35, 213
U.S. Bureau of Land Management (BLM), 117, 188, 194, 197
U.S. Department of Agriculture (USDA), 35, 43, 44, 144
U.S. Sheep Experiment Station (USSES), 43, **67**

V

vaccinations
  dogs, 89
  donkeys, 202
  llamas, 179–80
Valvasor, Janez Vajkard, 135
vegetation and electric fencing, 29, 32
verbal corrections, dogs, 61, 62, 70
vicuñas *(Vicugna vicugna)*, 163, 164
vision, donkeys, 189
visitors, dogs accepting, 60
visual deterrents, 33–35, *35*

vital signs, llamas, 182

Walton, Murray T., 190
water
    dogs, 87, *87*
    donkeys, 202, 207
    llamas, 178, 183
weasels, 18, 20
weight and aggressiveness, llamas, 166
weight of donkey, measuring, 205
whipworms *(Trichuris)*, 89, **89,** 180
white line disease (seedy toe), donkeys, 203, 208–9

Wild Free-Roaming Horses and Burro Act of 1971, 188
wild guanacos *(Lama guanicoe)*, 163
wire mesh fencing, 27, 28–29, *28–29*, 30–31
wolf-dogs, 11
wolf teeth, donkeys, 204
wolves, 3, 5, *9*, 9–12, *12*
wood chewing by donkeys, 211
Woodruff, Roger, 43
working livestock guard
    dogs, 39, 40
    donkeys, 190, *190*
    llamas, 164–67

X-style halters, 174

Youzhak (South Russian Ovcharka), *102*, 152, *152*
Yugoslavian Mountain Dog (Sarplaninac), 45, *101*, 135, *137*, 137–38, 142, 153

zip line, 81
Zoo-Veterinary Institute (Bucharest), 142

Other Storey Titles You Will Enjoy

The Dog Behavior Answer Book, by Arden Moore.
Answers to your questions about canine quirks, baffling habits, and destructive behavior.
336 pages. Paper. ISBN 978-1-58017-644-6.

Fences for Pasture & Garden, by Gail Damerow.
Sound, up-to-date advice and instruction to make building fences a task anyone can tackle with confidence.
160 pages. Paper. ISBN 978-0-88266-753-9.

Keeping Livestock Healthy, by N. Bruce Haynes, DVM.
A complete guide to disease prevention through good nutrition, proper housing, and appropriate care.
352 pages. Paper. ISBN 978-1-58017-435-0.

Real Food for Dogs, by Arden Moore.
A collection of 50 vet-approved recipes to please your canine gastronome.
128 pages. Paper. ISBN 978-1-58017-424-4.

Small-Scale Livestock Farming, by Carol Ekarius.
A natural, organic approach to livestock management to produce healthier animals, reduce feed and health care costs, and maximize profit.
224 pages. Paper. ISBN 978-1-58017-162-5.

Storey's Guide to Raising Series.
Everything you need to know to keep your livestock and your profits healthy. Latest title is *Meat Goats*. Other titles in the series are *Rabbits, Ducks, Turkeys, Poultry, Chickens, Dairy Goats, Llamas, Pigs, Sheep,* and *Beef Cattle*.
Paper. Learn more about each title by visiting *www.storey.com*.

These and other books from Storey Publishing are available wherever quality books are sold or by calling 1-800-441-5700.
Visit us at *www.storey.com*.